PRACTICAL LEADERSHIP SKILLS FOR SAFETY PROFESSIONALS AND PROJECT ENGINEERS

PRACTICAL LEADERSHIP SKILLS FOR SAFETY PROFESSIONALS AND PROJECT ENGINEERS

Gary L. Winn, PhD

CRC Press
Taylor & Francis Group
Boca Raton London New York

CRC Press is an imprint of the
Taylor & Francis Group, an **informa** business

CRC Press
Taylor & Francis Group
6000 Broken Sound Parkway NW, Suite 300
Boca Raton, FL 33487-2742

Printed on acid-free paper
Version Date: 20150921

International Standard Book Number-13: 978-1-4987-5822-2 (Hardback)

Library of Congress Cataloging-in-Publication Data

Winn, Gary L., author.
  Practical leadership skills for safety professionals and project engineers / author, Gary L. Winn, Phd.
    pages cm
  Includes bibliographical references and index.
  ISBN 978-1-4987-5822-2
  1. Engineering--Vocational guidance. 2. Leadership. 3. Project management. I. Title.

TA157.W54 2016
620.0068'4--dc23   2015035615

Visit the Taylor & Francis Web site at
http://www.taylorandfrancis.com

and the CRC Press Web site at
http://www.crcpress.com

Contents

A personal message to the safety and engineering professionals of tomorrow

Carl W. Heinlein is the past president of the Board of Certified Safety Professionals and was recently honored by the National Safety Council (NSC) with a *Distinguished Service to Safety* award at the NSC's 100th anniversary convention. Carl works as an insurance consultant with ACIG Insurance and worked earlier as safety and health director for Associated General Contractors (AGC) of America in Washington, DC. He is currently chair of the West Virginia University's Safety Management Graduate Program Visiting Committee, where his years of experience and knowledge have greatly benefitted the graduate students at WVU.

As I look back on the 20 years of my environmental safety and health (EHS) career journey, I am aware that I am still surrounded by some of my original educators, colleagues, mentors, plus union friends and management alike. If there is one consistent theme expressed by these individuals over my two decades in safety, it is that each has urged me to look beyond the traditional compliance-based programs and beyond merely management of people and go to leadership and leader development. I took that advice to heart, and as a direct result, I have been asked to be a leader and a trusted voice for both management and labor and for professional organizations. I take this responsibility and obligation very seriously.

My mentors and friends have been correct in their advice about moving away from simply management and toward leadership. Management is the theme of yesterday and leadership is the theme of tomorrow. Students and young professionals are justified to think that big change is coming. This is true for safety professionals and young engineers as well.

For example, only recently, we have received some great news about our profession. A survey has shown that the EHS profession is one of the strongest career choices for the next decade, with two jobs created for every college graduate. Even if the number of graduates in the field expands rapidly, the EHS profession is still going to fall far short of filling all of the open positions, and that trend will continue for at least another decade. And, while the EHS profession is a rapidly growing career path with opportunities opening every day, engineering continues to fill 5 or 6 of *Wall Street Journal*'s top 10 career fields year after year.

This book speaks to each group of young professionals, safety professionals, and engineers. Why? My experience has shown me that their jobs are different, but greatly intertwined. What one does influences the success of the other. As you'll read soon, both groups of young professionals are involved in decisions where people's lives are at stake.

As the EHS profession continues to expand rapidly in opportunities, so does it expand in expectations. More than other fields, all current and future EHS professionals will be asked to be generalists with an extremely broad base of skills. EHS professionals, more than people in most other career paths, must interact with labor, management, shipping, receiving, operations, and with the crafts and the general public. They will interact with the news and many types of media; they will interact with corporate leaders and academics. They will network extensively in and out of their own location. They will travel broadly. These young professionals must be able to grow and control their career in a predictable way and must learn to adjust rapidly.

A term that is commonly overused, yet holds very true in the EHS profession, is *multitask*. EHS professionals and engineers will do this on a daily basis, maybe hourly. They will learn the basics here in Dr. Winn's book.

I strongly recommend that young people choose to work for an organization that looks forward to encouraging its youth to become active in professional organizations and industry associations. These organizations and associations develop not only your professional skills and allow their members to network, but they also foster your ability to develop into leaders in industry and leaders in professional organizations.

I also encourage young EHS and engineering graduates to focus on continued education and training through professional certification and designations. This builds credibility for the pro, strengthens the field generally, and provides opportunities to take on additional leadership positions as they became available. Get a master's degree or a second master's degree; get an MBA; get an accounting degree. Stay active in the academic fields as well as in your professional organization. Never stop learning.

EHS and engineering professionals can and should extend outside of the work environment. In this way, our best minds help provide safer communities through involvement in civic groups, local police, fire and emergency associations, schools, scouts, and so forth. We probably haven't encouraged growth in civic areas as much as we should have to this point, but doing so helps take our unique skill set to other people who can benefit just the same as our employees.

A final important message to those who have chosen the EHS or engineering field: You will become a success when safety becomes a core value—not just a priority—with that professional's company or organization. When values are integral within a company's or organization's culture and when values drive decisions, priorities often change. People are energized, and maybe contrary to belief, our tasks actually become easier, not harder.

The future is bright for the EHS profession, in fact, brighter than ever. More companies and more organizations are finally working to integrate the EHS function broadly and to embrace EHS values and support the difficult decisions that EHS professionals often must make. What a gratifying career to be involved in, as the EHS professional partners in protecting people, property, and the efficacy of a company.

I want to thank Dr. Winn for finally writing his book about leadership and leader development simply because we have talked about this for over a decade. More than before, the timing of talk about leadership is opportune as the EHS field is growing rapidly, as companies truly embrace the goals of risk management and safety, and as we turn an important corner and move from merely managing the safety or engineering function to leading them. Speaking for many, I have valued his friendship and appreciate your guidance and mentoring of a couple thousand of West Virginia University's best and brightest EHS professionals that are working in a variety of industries worldwide. They make a difference every day. So can you.

Thank you and good luck in your careers.

Carl W. Heinlein, CSP, ARM, CRI

Top companies want leaders more than anything else

Eddie Greer is the director of business development for the Board of Certified Safety Professionals and is the past society president of the American Society of Safety Engineers, where he is a professional member and received the society's highest honor of Fellow in 2007. He retired after 31 years of service with Brown & Root/KBR, serving in safety and health leadership positions throughout the company in areas of SH&E. He says this about the book you are holding:

All you have to do is visit any bookstore, whether specialized or even at the airport, and you will find a plethora of books on leadership. Look at *The New York Times* list of best-selling non-fiction books and you will find well over half deal with leadership and self-improvement. One thing in common with just about all of these publications is that they agree that leadership is different from management. It has been my contention for many years that you manage "things" and you lead "people."

Leadership principles, especially for the up-and-coming young SH&E professional, is a soft skill usually not covered in an educational environment. As young professionals, you may not consider it as important as your technical skills or give leadership much thought, but you will find that obtaining and utilizing leadership skills will become a key component if you are to be successful and efficient.

Dr. Winn starts at the beginning with discussions of what it means to enter a profession. That must come before any discussion of leader development.

But one thing is certain, there is no quick path or silver bullet in leadership development. It takes many years of study, experience, and mistakes. A degree simply gets you in the door, and from there, you must work to obtain, sustain, and develop the skills to be an effective leader. Being a leader is a journey and not a destination. Even the mightiest of the oak trees die when they stop growing. The same can be said of being a leader; growth and learning are synonymous with success.

I have taught classes on leadership for many years to all levels of participants: from young professionals who are new to the business world to so-called leaders well established in the work environment. The need for effective leaders is still very much evident, and more so today than ever before.

The good news is that leadership can be learned. This book will explore many aspects of what makes an effective leader. Dr. Winn has researched many avenues, including both the industrial and military approaches to leadership, to develop this book and help begin your journey to leadership and "making a difference." That's precisely what safety professionals and engineers do that distinguishes these important career paths from all others.

Chapter 2, "Self-discovery Comes First," ties in very well with part of my definition in that knowledge is not enough; you must understand where you are with regard to leadership and understand what is needed to get you where you want to be. To visualize yourself as a leader, some key questions need to be answered: Who do I want to be like and equally, or even more important, who do I NOT want to be like? Finding a great leader who you can use as a mentor will help your development. You also need to scrutinize yourself for strengths and weaknesses. What steps are you currently taking to help yourself grow and are you actually doing something proactively? Finally, you can actualize your current situation. Are you implementing proactive changes to improve your leadership characteristics? Recognize the things that you feel are encouraging and the things that are tearing you down. Develop an action plan to keep you positive and on track.

In Chapter 6, "Culture, Safety, and Engineering," leadership plays a key role in developing a sustainable culture. I have a good friend, Brad Giles with the URS Corporation, who states it very well: "Leadership drives the culture and culture drives safety." Dr. Winn approaches culture not as an abstract concept but from empirical research emanating from Edgar Schein, the foremost thinker on the topic. If you say "culture" in connection with leadership, it probably began with Schein.

One important part of this book is Dr. Winn's discussion of the "depleted environment;" what a safety professional or engineer can do when faced with an unsupportive climate or a huge corporation where nobody else seems to care about leadership. Can the committed young professional leader really do something to create a miniculture? Winn says yes, by following a research-supported recipe he has developed on his own: beginning with an honor code (he suggests a well-known sample) and moving to "values congruency" where workers and managers *are* the ideal, *know* the right thing, and then *do* it without fail or apology. Finally, in this simplified model, safety pros use storytelling, nonmaterial rewards, and personal courage to keep the system refreshed even when there is no support from top management. These aren't new concepts (Dr. Winn credits its original authors), but his application is completely fresh. This little section in the book may end up being its most important.

The real meat of the book is found in Chapter 10, "Crisis and Noncrisis Leadership Models." As you grow as a leader, you will have victories and defeats. How you handle both of these will tell a lot about how you will be as a leader. Trust me in the fact that there will be times when you wish you weren't in a leadership position. However, when you experience victories and start to make a difference with your people, the company, industry, and the profession, all the hard times will melt away. It's how you handle those hot water situations that makes you a stronger and much better leader.

Dr. Winn has provided you with the very latest thinking from industry and the military, researchers and philosophers, and the occasional tank driver. In many cases, you'll note quickly that many of the references are less than a year old as I write this. Your challenge is to recognize the opportunities and apply what you have learned to help influence the people who are your followers. Your goal should be to become an effective leader and serve as a mentor using these fresh approaches.

Good leaders take people where they want to go; great leaders take people where they don't want to go but need to go. This book will help you become a great leader. I am confident of it.

Sincerely,
Eddie Greer, CSP, OHST

Preface

Rewarding careers with big potential consequences

To preface the material in the book, I want to share three observations that underscore the need for a practical book on the art and science of leadership and leader development. The profession of leading and controlling a big safety operation has huge rewards, including advancement both vertically (same industry) and horizontally (different industries). Similarly, project engineers continue to be in huge demand in construction, petroleum, and manufacturing, and more. But unlike most service industries, our chosen career fields can have serious consequences for personal, property, and business continuation when things do go wrong. That's the whole point of this book: My experience and data suggest that both graduate and undergraduates in these fields could be better prepared for the "big time" not only in terms of professional development but also in terms of being ready to be a leader and to develop their own pipeline of subordinate leaders. In this preface to the material that follows, I will justify my concerns.

First, I am concerned about our future safety and engineering leaders as they begin to enter the workforce. In many ways, they are less prepared than they would have been two decades ago, and I discuss my own research later in these pages to justify my contention. On the one hand, recent graduates are better prepared on technical content than any time in history. They can probably calculate time-weighted averages and Laplace transforms in their sleep.

But some years back, I began to sense a change in incoming students. I decided to conduct surveys to verify my suspicion that these were tangibly different students. I found that most of our college-aged students—the Millennials as they are called—have not worked over summers or during the school year and so they don't know how an office operates or how a memorandum is composed. Basic office protocol and etiquette seem to be a mystery to them.

In addition, they have not traveled widely and apparently don't even want to, which means that they are not as ready to understand diversity and cultural nuance as they might think. My work shows that, increasingly, they don't read newspapers or novels or biographies, even eBooks on a Kindle or Nook.

These graduates vaguely understand that they are entering a profession, but they don't know what a professional really does. They have little concept that they will be faced with ethical considerations far more demanding than their friends in low-stress, low-risk careers.

I tell them only half-jokingly that they will miss important business opportunities if they don't read or travel enough to strike up a good conference dinner conversation about shale gas or smart materials, for example. What are the rippling effects of energy prices on miner safety? Can the fast-paced world of unmanned aerial vehicles (drones) impact worker safety? There are no concrete answers for these and there are a thousand more questions drawn from a news website or morning paper, but a well-prepared young professional on the way to leading his or her department and subordinates will live in a smaller world year by year and should want to ferret out answers to help them lead.

Unfortunately, I don't think professors challenge students very much to read, to travel, and to grow, and as a result, graduates don't seem to understand how interrelated the daily news, current events, travel, and their careers really are. It really matters that our students become more globally ready and can think about the rippling economic effects of far-reaching political climates as they affect worker safety and engineering best practice (for example, safety-through-engineering design means that safety pros and engineers need to know what the other does).

Recognizing these "missed opportunities" for work, travel, and reading, I see that young people in my classes and at the threshold of their careers are more siloed and less global than ever before, but it's nothing that can't be overcome. I show how to address these missed opportunities in this book.

Among the following chapters, we're going to start with the basics: positive preleadership activities, and one of the things I will suggest is continuous reading of *nonwork, non-academic-related* material. And while I admit that my suggested reading lists are based in Appalachia, where (by my count) we have nine universities within 200 miles that are preparing safety and engineering students, a regional reading list could be just as easily for the Southwest, New England, or any other place a leader developer takes an interest to do so. I urge professors who may adopt this book to create their own reading lists.

Second, I have noticed a trend in the last three to five years where industry and government are asking no longer for managers, but leaders of change. It's clear to me that leadership is the new buzz word for engineers and safety professionals. We have spent decades preparing technically qualified engineers and managers, but when it comes to preparing leaders to meet the new demand, I'm afraid academics aren't themselves prepared to teach about leadership or training subordinates to become leaders.

Let me support my point, first in the safety field:

- Simply searching the Internet for "safety leadership conference" brings up 47 million hits. That's right, *million.*
- For over 10 years, ASSE has been offering an annual "Future Safety Leaders Conference" for the specific purpose of safety students "becoming more effective leaders and communicators." The speaker for the 2014 executive session, Dr. Daniel Moran, spoke about "acting and demonstrating safety leadership, even in the face of difficulties," the very same topics and needs for the future as I discuss in these pages (http://www.asse.org/membership/student_fslc/).

It sure looks like the safety field is interested in teaching leadership. But what about engineering? Here's a fairly typical example. As part of a much larger strategic plan, the American Society of Mechanical Engineers surveyed 68 academic department chairs about communication, ethics, and leadership skills among their graduates. Only 20 percent of these academic chairs considered their students' skills to be weak. Yet Donnell et al. (2010) reported to the American Society for Engineering Education:

> Unexpectedly, a parallel survey of industry representatives found almost opposite results, with only 9 percent considering communication, ethics and leadership skills of recent mechanical engineering graduates to be strong and 52 percent of those same students to be weak. Given these results were gathered from 68 mechanical engineering department heads and more than 1000 engineers and managers [currently working in industry], a disparity clearly exists between the communication, ethics and leadership skills we are teaching to engineering students and what industry expects our students to know.

I have to conclude that while academic engineering departments think they are supplying leadership content (among other content), industry seems to think otherwise, and that leadership and ethics should be given more priority.

So, what about a daylong conference on leadership, or maybe a webinar? While well intended, not only are these conferences impossibly brief, but also under the surface, the material is most often merely collected wisdom handed down by an experienced professional or a big name somewhere. I will admit that good stories make easy reading, but unfortunately, the stories often pass for leader development material. I'm afraid they're just not the same. I further admit that stories, accumulated wisdom, and anecdotes have their place and I use them in this book, but they have to be balanced by corroborating empirical research and data on outcomes. That's the balance I try to strike in this book.

I know that this book is impossibly brief, too, but it does attempt to set the stage for a lifetime of leading and developing subordinates into leaders. It starts with a *conscious decision* to become a leader and not just an employee.

The organization of this book is part empirical, part anecdotal. Its chapter organization parallels the steps that a student or recent graduate needs to make to progress from preprofessional to professional to leader to trainer-of-leaders. Once the new graduate actively chooses this path for his or her career, this book offers a rapid fire way to move forward in a world where the clear expectation is "leader."

Other career paths have the clear expectation of "leader" for their young people, and early in the course of preparing these materials, I noticed the copious organizational research produced by military behavioral scientists often teaching at our service academies. One particularly good example is a book I reference a few more times later in this pages, *Leadership in Dangerous Situations: A Handbook for Armed Forces, Emergency Services, and First Responders*, written by Patrick Sweeney, Michael Matthews, and Paul Lester, all with doctorates in the behavioral sciences and all having taught at West Point.

The more I read, the more I realized that what the military professionals do and what safety and engineering professionals do on the civilian side are much the same. We both work sometimes under intense pressures and we have to be ready for the volatility and challenges in our respective fields wherein bad decisions can be deadly. You'll see that I have sampled greatly from military scientists, and as this book preparation winds to a close, I wonder why the important textbook authors in our field seem to ignore military leadership science.

Third, I wrote some important material specifically for safety professionals and engineers already in the field and working. I decided to venture into a couple of places where others have not trod, so far as I know.

For example, what if after two years in the company, you realize that upper management isn't interested in developing leaders in-house. What if the CEO doesn't care about values-consistent safety behavior in what I call an isolated and "depleted environment?" My answer is that the leader or leader-trainers must actively take the initiative, even alone. They choose a code of behavior supported through values that they select for themselves; they use tested methods of changing a culture to support actions that are values consistent. They consciously support behaviors that, while they appear "safe" in the traditional sense, are really only actions consonant with a basic duty to look out for each other. And all of this can happen from the bottom up if a leader developer *chooses to do it*, as I discuss in a later chapter.

Another area important for those already on a career path is what to do about "toxic leadership," which occurs almost predictably when management practices favor the status quo. Why is this important? Because, contrary to the workforce of a couple of decades ago, our new safety professionals and engineering careers don't have work experience to help spot destructive leaders. I offer help identifying and fighting it.

A third example of material useful to the recent grad or for any young leader who may work under really hazardous conditions is the work I cite from established researchers on how to handle "the death (fatality) event" in an organization. Of course, we hope it never happens, but this is the business we are in, and I think it's best that young employees are prepared.

Later, I offer a chapter on gender advice for tomorrow's leaders because the Millennials have a different view of diversity than the older generation does, including me. To help with this chapter, I reached out to women already in high-visibility engineering and risk management careers. I have profiled some interesting work by Barling (2014) and also by Sandberg (2013) about problems and solutions in the gender arena, but I balance it by making sure we have current views from young women who share what it's like to work in the gas field or on a construction site. I think many readers will find this information insightful.

Finally, and for those already on a career path, I gathered suggestions about international travelling etiquette, knowing that more than ever, recent graduates will be working and living abroad. I asked a group of culturally diverse graduate students from a half dozen representative countries how to avoid embarrassing mistakes overseas.

A brief word about my approach to this book and to friends who will choose to become a leader

If I seem cynical or flippant on occasion, it's my way of expressing what I believe to be some particular unvarnished truth—and sometimes I am cynical or flippant in real life. My years of experience have told me not to take myself too seriously in the classroom or on the shop floor, so I don't. For purposes of this book, I wanted to have a conversation with the new graduate and not present just another lecture. In all earnestness, I didn't set out to write a stuffy textbook, and so I didn't.

I try to justify my work here with published research combined with my own experience in industry, which included a couple of stints as an engineer, and of course, my experience as a professor. I know that a student can spot a professor without real work experience a mile away and will shun him or her—I fully appreciate that.

I've tried hard not to embellish actual anecdotes I use from real industry, from real worksites, and from real people. I have eliminated any personal anecdote that did not deliver a clear message.

Finally, let me say that there is no rational way to treat in a single text the full array of topics needed to prepare our safety and engineering leaders of tomorrow in leadership or leader development. Besides, there will surely be better books written purely on the science and organizational research of leadership or purely on its maxims and wisdom. What I have endeavored to do is blend research on leadership with wisdom and anecdote and do it in a way that aims directly at young engineers and safety professionals.

While I don't claim to have covered every leadership source out there, I have distilled the best sources I could find for the special purposes I have set forth for this book as I have outlined in this preface. For those who *choose to lead*, here's a good place to start.

Thank you for allowing me to share with you what I have found.

Gary L. Winn, PhD
Industrial and Management Systems Engineering
The Safety Management Program
West Virginia University
Morgantown, West Virginia

Acknowledgments

Acknowledgment for encouragement

I thank the following people who have provided important moral support through the ordeal of research, data collection, and data reduction and, finally, writing this book.

Eddie Greer was encouraging even before there were words on a page. Amanda Crosby, for whom I am an academic advisor for life, encourages me in real time. Sarah Soliman, my engineering freshman and probably the future governor of West Virginia, has inspired me for a decade. Carl Heinlein, the likely future president of ASSE and always a cheerleader for leader development, offers me kind words regularly. Ed Youngblood doesn't stop trying to encourage my writing technical and creative material. Casey Brower encouraged my daughter to be a historian, but she ended up an engineer anyway; he has forgiven that and has been a kind and inspiring aura for both father and daughter. Kate O'Hara, Jennifer Worthington, Hillary Dean, and Jenny Fuller are a few of the top-shelf female safety professionals I am proud to have studied with me over the years. Greg Harrison showed me that even funny stories can carry a sobering message. "Rock" Roszak gave me wonderful first-hand insights on the riveting story of the B-52 crash involving West Virginia's own Mark McGeehan. Dr. Kevin Rider, Burdell Brock, Jody Gray and Pat, Dr. Christina Wildfire, Frank VanCleve, Dr. John and Joan Spaulding, Raymond Stockdale, and Dr. Ava Dykes have stood with me in some pretty difficult moments while this book was being prepared; they are my friends for life. Austin Lee Winn and 2Lt. Laura Dukens, two of the young engineers I am writing for, show me daily what honor and especially loyalty can mean to friends and family. I have enjoyed many-faceted discussions on military history, the Jesuits, and spiritualism with Dr. Jeremy Slagley for years, and I still do. My mom and dad, the end of our family's line of the Greatest generation, said for years that I ought to write a book to help my students, and so I did.

Acknowledgment for content and commentary

I would like to thank the following people who have provided important technical material for this book.

Dave Miller is a cherished family friend who showed me the life-changing qualities of servant leadership, which I highlight in the book. Despite my own best efforts, Jason Musteen brought new and meaningful messages about the 22-year-old George Washington's leadership to light. Ron Kasterman showed me new West Virginia historical vistas with his kind gift of *That Dark and Bloody River*, which should be required reading for students east of the Mississippi River. Fred Schroyer showed me why I'll never be the copyeditor that he is, although he did his best to train me. Fortunately, you can't sell books back to the bookstore at West Point, so Laura Dukens gave me a three-foot-tall pile of books and historical maps from her classes on leadership and history. These are books that I cherish. Bob Hayes, the former president of Marshall University, and his son Mark, an attorney friend of mine, have provided me with material and enthusiasm for West Virginia's sons and daughters. Amanda Fulk helped me to control my loathing of certain word processing software. Alexis Williams is another promising graduate student who provided personal support along with insight and encouragement. Tom Kolditz, Bernie Banks, Mike Matthews, Casey Brower, and Tom Merriwether were awfully patient in explaining how wrong I was to think I could somehow "distill" leadership into a few theories and maybe a single college course; I have grown immensely in their shadows. Josiah Grover is a talented professor specializing in historical weapons and why weapons technology in the hands of a brilliant leader is a key to winning in combat. Professor C. B. Wilson agreed that this book was a good idea and approved a sabbatical so I could devote time to writing it. C. B. has been the associate provost for academic personnel at West Virginia University for a long, long time and we are lucky to have him. Drs. Tara and Dan Hartley are epidemiologists at NIOSH in Morgantown, West Virginia, and provided important material about work-related stress, in particular, the Buffalo COPS study. They are good people and proud WVU grads. ASEE's Mike Burditt provided an endless series of constructive comments for which I am grateful. Taylor & Francis' Melisa Sedlar and Cindy Carelli provided useful commentary in the final stages of the book's preparation.

And to my safety and engineering students over a quarter of a century, thank you for letting me be part of your lives.

This bulleted abbreviated roadmap represents a time-tested, research-backed way to accelerate a recent grad or a new hire in safety or engineering.

Unit 2: Understanding leadership

Becoming a leader isn't a spectator sport: It's a participant sport and a lot more like rugby than billiards. Yet despite getting bounced around, I have found that the most satisfying and successful careers in engineering or safety will be built upon "giving to" and "giving back." *Giving to* means giving your best effort to your employer and acting consistent with your own core values. *Giving back* means actively preparing your own subordinates with the skills and values they'll need to replace you once in a while, or even permanently. I have found that giving to and giving back are life changers as much as career changers.

And they are characteristics of solid leaders, and leaders *choose* to do them.

In this segment of the book, we discuss crisis and noncrisis models of leadership and how today's foremost researchers in organizational behavior view these topics.

Unit 3: Applying leadership fundamentals

I have placed graduates and mentored early career safety and engineering professionals in careers for decades and I noted for a long time that their training and early career progress rarely consider leader development. There are no books available on the application of leadership principles, and so I decided to write one myself.

Why not just write my own textbook, they said. Sure, I said, I've got a couple of free weekends. Boy, did I underestimate the task…

I have tried hard to find and incorporate as many actual case studies to demonstrate the principles I illustrate both in safety and in engineering.

One important part of this module is a new model for experiential training, something we do in safety and engineering all the time, but for certain kinds of training, there's a way to merge the acquisition of knowledge and skills with leader development. And there is a good bit of research to support it, too.

Unit 4: Fine-tuning leadership applications

This book is an effort to combine the art and science of leadership in a common-sense way; you'll soon find that my tone is conversational and the presentation is not going to exhaust every textbook out there. Rather,

I'll cover the material that, in my experience, will be germane and save time exploring for "what works."

In the basic books on leadership, there are no discussions of "toxic leadership," or what happens when good leaders use their nominal positions to bully or serve themselves. I did not see anything about how a young professional can actually create a microculture in a climate where upper management does not seem to care about leader development. I found nothing about some of the unfortunate but likely circumstances that a new safety professional might discover when a colleague or direct-report employee is fatally injured on the job.

I cover those and a couple more areas that I found where my students could use a jump start: basic office protocol and business etiquette.

section one

Choosing personal development

chapter one

Why leadership and why now?

This chapter represents a "needs assessment" that suggests that while the industry at large is crying out for safety and engineering leaders, graduates and young people early in their careers simply haven't been exposed to what it means to be "a professional," much less a person who will very soon have the responsibility for subordinates. This chapter sets the stage for the reasons a book about leadership is needed.

I have absolutely cherished the time I have spent outside the classroom talking to my students about their dreams and hopes for the future. Some conversations are a few minutes and some can go on for hours. We have discussed career paths of course, but also things that matter to the students: knife making, the French and Indian War, motorcycle trips on back roads, horsemanship, the Jesuits, Barrett 50 calibers, and even golf. (Ok, not golf. I draw the line there.) I have taught engineers as freshmen who have gone on for a doctoral degree and who are themselves professors. I have taught safety students who have risen to the very top of American industry, and engineers who have hundreds of subordinates working for them. But in talking to them, caring about them, "adopting" a few who needed it, I have learned a lot. These seasoned pros tell me that we are producing technically competent young people but immature in the ways of professionalism and leadership.

The students and their interests are all over the board with one exception. *Universally, they have a desire to make a difference in people's lives.* They are motivated to help, care, and protect. What I do is help create a plan of study that gets each one to their academic and career destination. They provide the more important part: motivation to work hard in school and then at work, and altruism to preserve and protect.

My part is small, really. In some ways, I am only a cheerleader.

Young people are primed to take on careers to preserve and protect not only because they are technically competent but also because they are altruistic. And despite wars in 20 countries going on as I write this, they are optimistic about the future.

But despite the optimism and enthusiasm and stunning technical capacity, they are not quite ready to walk in and begin to change the world, because unlike their predecessors of 20 years ago, graduating engineers and safety students have limited preparation in travel, extensive reading, and work experience. In fact, the Donnell study I cited on p. xxi

3

from 2010 is exactly what I have found in my own research here at West Virginia University (WVU); the university—whether chairs or the students themselves—report that they are ready to go, ready with technical skills, communication, ethics, and leadership preparation. But industry says the opposite (see Donnell, 2010, or search the recent proceedings of the American Society for Engineering Education [ASEE]).

My position is that our technically prepared youth are not quite ready to enter a global and fast-paced workforce, but there is something else endemic to these two career fields. I am particularly concerned about our youth going into career fields where the consequences of a bad decision may be fatal. At a time when industry, government, and even the nonprofit sector are asking for leaders, the academic community really isn't ready to respond. Not yet.

The "Millenial generation" is a mix of challenge and opportunity

Are youth today really different? How so? Let me set the stage, but before I do, let me say clearly that I am not complaining about their preparation or condescending. I am merely saying that they have missed opportunities to learn about humility, courage, and self-awareness. Four years of my own data support my hunch: These students have missed important opportunities to prepare themselves for the big leagues of budget, staff, travel, and global responsibility.

The most effective path to leadership involves learning vicariously from not only others but also a wide range of reading and even travel.

When I ask incoming graduate students and engineers in formal surveys if they have worked over summers or in high school, fewer and fewer say "yes." When I ask them if they have ever managed anything at all, even a swimming pool or fast food shift, the answer is almost universally negative. When I ask them about travel or broad-based reading, they say they do not and don't really care to. When I ask them what they read just to make them an interesting conversation partner among colleagues, they are silent. They do not read much outside the classroom. Providing these results later in some detail is not meant to be condescending or to build a negative generalization. On the contrary, they stand as a platform from which to build. These data will be our starting point, then.

The leadership challenges are real: Even now, some graduates are going directly overseas immediately after they graduate; all are managing people and big budgets; and they are directly responsible for people's lives. At age 22 or 23, these missed opportunities for growth, travel, work experience, and even making good conversation have huge potential consequences. It will take them a year, maybe two years, to catch up with people the previous generation called generation X who were, honestly, more prepared. Again, I don't say this to be unkind. It's merely that professors need to help students catch up as quickly as we can on the soft side of their preparation. Their technical preparation is never in question, only these soft skills.

On the positive side, young professionals in this generation are driven by high motives, and I hear that in talking to them regularly. They are eager to start making a difference and they are more technically qualified than ever before. But they do not have experience under their belts yet; most don't even read daily newspapers or weekly news journals. These students are much, much different from students a short generation ago, who were usually already employed and midcareer when they started the safety academic program.

Unfortunately, academic programs in safety and engineering have spent precious little time on the study of organizational values and how those contribute to organizational culture. Academic programs don't spend much time on leader development, ethics, or company protocol. The same programs don't even differentiate between leader and manager. Most graduates are technically qualified to start work in the strict sense, but they are not ready to become leaders of tomorrow, which is precisely what industry and government and even the nonprofits are asking for (Figure 1.1).

A prominent book from the last decade-plus describes the Millennial generation (also commonly called the Y generation) that bounds young people born between 1982 and 2002 (*Millennials Rising*, Howe and Strauss, 2004). This book inspired many others that describe this generation as overprotected but still eager to learn. Two other books in this category are *Y in the Workplace: Managing the Me-First Generation* by Nicole Lipkin and April Perrymore (Career Press, 2009) and *Not Everyone Gets a Trophy: How to Manage Generation Y* by Bruce Tulgan (Josey-Bass, 2009).

The upshot of these books is that the Millennial generation is different enough to require special methods of managing and motivating them. What has not been said yet is this: The first of the Millennials started their working careers in about 2002, and they entered middle management in about 2012. Millennials are probably overprotected but fundamentally altruistic and seek to work by conservative and strongly held values. But because they have missed opportunities to work and learn about them first hand just as they enter the part of their careers where they set policy and are expected to step into leadership safety and engineering positions, don't we owe them some kind of formalized, structured presentation of just what a leader does? If we expect them to "move the needle" or directly intercede and impact culture toward values-driven safety and engineering responsibilities, ought we not talk about what it means to move the needle?

Figure 1.1 The oil and gas industry has opened many jobs for both safety professionals and engineers since 2010. (Courtesy of Jennifer Worthington.)

The Millennials represent a 20-year span of young people born after generation X. These young people have solidly entered the workforce after college, or about 2002–2004, and now 10 to 12 years later, they are entering leadership slots. In terms of size, the Millennial generation is big, very big. In fact, it is the biggest generation to come along in 50 years, and the reason seems to be because the Baby Boomer generation preceding it delayed childbearing for a longer period than in the past, with adults midcareer themselves and having children for the first time in their 30s and 40s.

The Millennials are, according to Howe and Strauss (2004),

> Wanted. Protected. Worthy. Thus did the heralded
> Class of 2000 arrive in America's nurseries and cribs
> [starting in 1982]. Soon a glossary of mainly positive

adjectives would describe them. From conception to graduation, this 1982 cohort has marked a watershed in adult attitudes toward, treatment of, and expectations for children. Over that eighteen-year span, whatever age bracket those 1982-born children have inhabited has been the target of intense hope, worry, and wonder from parents, pollsters, pundits and politicians. (p. 32)

Howe and Strauss say that the Millennials have these seven characteristics that distinguish them from generations before or since:

- They are *special* and parents are demonstrative and lavish with praise ("every kid gets a trophy" no matter their skill level).
- They are more *sheltered* by parents and governments than at any time in history from consumer products (tamperproof bottles, air bags) than any generation before or since.
- They are *confident and optimistic*; they are perhaps overconfident and sometimes they boast.
- They work best in *teams and groups*; they are interconnected 24/7 by a handful of technologies and are thus extremely collaborative.
- They see themselves as *achievers*; they go to class; they have pretty good moral compasses.
- They are *pressured* by parents and teachers to take part, to apply for yet one more scholarship or college. They will tell you the pressure wears on them.
- They offer and hold fairly conventional values systems that are similar to those of their parents, many of whom openly rebelled in the 1960s at their own parents' value systems.

On the very good side of the ledger, these same kids are driven to be altruistic, caring, if a little independent. Despite the criticism that they have been pampered by "helicopter parents," they are civic minded, they volunteer their time, and they have more conservative and traditional values than did their Boomer generation parents. They want to preserve and protect.

Millennials: Who better to have as a safety professional or project engineer?

Maybe Millennials are, in fact, ready to preserve and protect, but what they don't have is experience. Yet, we academics and current industry professionals recognize that despite being pampered and despite most of

Generation name	Born	Started work	Entered midmanagement
The G.I. generation	1910–1922	1930–1942	1940–1952
The Baby Boom generation	1946–1964	1966–1984	1976–1994
Generation X	1965–1981	1985–2000	1995–2011
The Millennial generation	1982–1999	2002–2019	2012–2029

Figure 1.2 The Millennials have entered middle management in recent years.

them having high motivation to improve people, at least half of them are still in school. These are the last part of the Millennial generation in late high school and in college, students who are receptive and can be reached with novel ideas about values-centered leadership. The way I see it, they are motivated and want to be leaders, and they are confident enough to carry out their high aspirations. Let's help them while they are still at the university or else early in their careers (Figure 1.2).

But what about those who are out of school, on the front end of the Millennial generation who are entering middle management in the mid-part of this decade? They are entering positions of policy and authority in industry, educational institutions, and the nonprofits. They will begin to set accountabilities, determine marketing directions, and for our part, become chief safety, health, and environment (SH&E) officers, or else senior engineers. The former will have achieved their Certified Safety Professional (CSP) and the latter their Professional Engineering (PE) certifications.

Are the Millennials ready to "move the needle" and become the safety and engineering leaders we are all asking for (Figure 1.3)? Are they ready to become change agents? Unfortunately, I think not, but we can change that.

Wouldn't it be nice if we could offer those motivated, altruistic, driven youth who are between still-in-school and midcareer advice on establishing a values-driven system of leadership and leader development? That's the central purpose of this entire book.

My data, presented later in more detail (see Winn, Giles, and Heafey, 2012, and Winn, Williams, and Heafey, 2013), represent small samples taken at only a few universities, but in each survey, the students are enrolled in majors where they will encounter hazards and where they will be expected to lead people in hazardous situations. The data are presented

> If we expect the Millennials to "move the needle" as they become the professionals and leaders of tomorrow, we ought to talk about what it means *to become* a professional and leader.

Figure 1.3 We expect Millennials to "move the needle," but they must be prepared to do so.

more fully later in the book, but for now, let me skip to the conclusions about why we need to care about the Millennials. We have the right people and their information about how to create values-driven, authentic leaders, even when the uber-organization does not support leader development. I show in the book that motivated and knowledgeable leaders can make a difference, and why they should do so.

The opportunity is to take advantage of this generation's motivation, altruism, and hope. The challenge is to prepare the fundamentally unprepared members of the same generation.

Shouldn't we also want to move the needle toward values-based safety and leadership?

I'll never forget being overwhelmed—I mean, stopped in my tracks—the first time I visited the bookstore in Thayer Hall at West Point. There were racks and racks of textbooks on leadership. I saw books by military writers, leadership books by former presidents, books on leadership by European military historians, and books by industry experts and Rhodes scholars, all about leadership and leader development. As I recall, there was an entire aisle in Thayer Hall devoted to books on leadership. But, undeterred by the sheer volume of books, I bought an armload and dug in to see what I could boil down for my students. Later in another visit, I bought a second armload of books on leadership at Thayer Hall. This was going to be pretty easy, or so I thought. Four easy steps: read, digest, write, done.

Yes, I originally thought this leadership stuff was going to be pretty easy. I admit it freely. Professors are sometimes pretty naïve. Ok, they are *often* naive. I had no idea there was so much good material on leadership, ethics, and protocol available outside safety and engineering.

Since my visits to Thayer Hall, I've bought more books outside of West Point, and later, more books still on Amazon and everywhere else, including used book stores. I've scoured the country for books out of print and obscure but germane research manuscripts.

On a pair of side trips, I realized that West Point seemed a logical place to gather research and material; after all, many of the top management consultants in the world consider West Point to the best leadership training location in the United States. I was able to speak to some of West Point's best and brightest as I gathered material.

Future safety leaders and engineers take note: Life isn't going to be easy. Inside the first month, you're going to wish you were back taking finals and doing labs where it was pretty stress free and, certainly, the hours were shorter. Instead of easy, job life begins quickly and ramps up startlingly fast. One recent grad found himself with a surprise promotion, which meant managing a team of a dozen environmental health and safety (EH&S) staff at a tire factory with two unions. He was responsible for over 3000 employees in Ohio, all inside the first year of employment.

In the last three years where this trend has been growing quickest, I have had recent engineering and EH&S graduates placed overseas in the Middle East, Africa, and the United Kingdom right after graduation.

My own son, a young mechanical engineer, doing piping design work, was on his job for exactly two days before his boss tossed him the keys to a rental car and slapped a credit card on the desk. It was his introduction to the big league: Go forth and do good, manage budget and staff. Show the contractor you are a professional (and that you know your job) and that you are also an interesting person (to cement the business relationship). My own son's experience proved my point: For him, it was meeting the big time inside of two days on the job, and it may happen to you too.

I am absolutely convinced that young people ought to be doing everything they can do to see themselves as future professionals and, ultimately, leaders. They need to know what professionals do and how they must act. They need to learn what differentiates good managers from good leaders because their expectations have never been higher. And they need to learn quickly.

I am betting nobody has mentioned in class how fast real life hits you.

Regarding "interesting," one of the first things I will impress upon anyone thinking about 30 or 40 years as a professional is the simple need to become an interesting person. People will seek you out soon enough if you are not stuck at the water fountain talking about hockey or, God forbid, golf. When you stretch your horizons, you will find kindred intellectual spirits everywhere. New networks of challenging, gifted people will open up for you, and they will seek you out, I promise you here and now. Interesting people have a kind of radar for other interesting people.

Maybe you had a great conversation on the airplane or the train. You can lock down a promising business contract this way because the negotiations began with a great conversation. At a bare minimum, you can expand your professional network by becoming merely interesting.

Most of these fateful conversations begin with ideas outside of work. I can prove it a hundred times over through experience. And here's a tip, the first one in this book: You can be a lot more interesting starting today if you just replace a lot of your sentences that end with periods with sentences that end with question marks.

You can thank me for this later, but I actually heard this from Frances Hesselbein, the well-known author on leadership topics, and she was right.

You need to be a *professional* first, and then you need to work hard to become *interesting*. And I may as well get this bias out on the table now because I have biases generated over a long time of experience. I most certainly do not mean the following things qualify as "interesting" topics: sports, especially football and golf. It's a brief list of things to avoid. Why? You lose people regardless of gender who want to talk about more exciting ideas. Go from there.

If you are a young safety professional or engineer and meet me at a conference some day, you'd better be prepared to discuss something interesting and challenging and thought-provoking that you read recently outside of class and outside of work. If you're going to have dinner with me at that conference, and please understand—I hope you do—know something about the French and Indian War, or economic theory or manufacturing processes, Keynesian economics, pop culture, or something you read that morning in the *Wall Street Journal* (*WSJ*) or *The Economist*. These will go a long way at igniting interesting discussions. Every business leader wants to be challenged in this way, not just professors. You will be a far better professional and a much more effective leader if you become interesting at the very outset: *interesting* because you can share an intelligent discussion, but *interesting* because you can speak to global issues which will impact your and intersecting fields. Make sense?

Back to my epic quest to write a book.

I embarked on a trip up and down the East Coast interviewing experts and attending conferences on leadership, talking to industry up-and-comers and members of the Greatest generation. At the end of it, I realized that I had been fortunate and lucky, honestly, to have discussed leadership and leader development with a dozen of the country's top experts. Some of these people had written the very books I bought in Thayer Hall at West Point. Apparently, I am not alone in thinking West Point's bookstores had excellent leadership materials.

As the fourth year of work on this book comes to a close, I have tried to condense the essence and thinking of a lot of people about these topics upon which so much is written. I have tried to condense for safety and engineering students what a leader does, and why this is so much more important now than even a decade ago. That's the essence of this book. It's a much tougher job than I anticipated. Maybe I have stretched my horizons and become a more interesting person along the way. Something tells me I have.

On a side note but an important one, I discovered that some material posing as leadership theory, ethics, and protocol is wisdom of the ages disguised as research. I try to point it out on these pages when I see it, but still, empirical research just doesn't cover all of the bases for information we can use. Sage wisdom—good advice—is important, too, and so I present the good stuff when I find it, even though my training looks down on it.

But we need to be skeptical of merely advice from crusty experts, even the sagest and time-tested advice. It does not substitute for a data-based, empirical treatment of values-based leadership and authentic leader development. There are no shortcuts; there are no 12-step programs to leader development even though there are some books whose authors would have you believe it. In reality, and as I was warned repeatedly during my interviews, there are no quick fixes. We need to ask for the data and inject them into our veins. Sage wisdom masquerading as research will not satisfy the craving for knowledge.

Figure 1.4 presents a way to think about how to balance leadership research with leadership anecdote.

Type of material in the literature	Advantages	Difficulties	Estimated percentage of the leadership literature at-large
True experimentation on organizational behavior and leadership Example: Matthews' work on comprehensive soldier fitness (see Chapter 12)	In a true experiment on a leadership theory, an investigator can "prove" a theory. Cost is high because many groups and controls must be put in place.	1. The investigator can't assign subjects to test-groups in studies on organizational behavior. 2. Control groups (no treatment) can't be formed, either.	Less than 1 percent
Quasi-experimentation on organizational behavior and leadership Example: Collins' work using financial records as the main factor in organizational leadership (see Chapter 10)	Proof can be established statistically to a high degree of probability. Leadership Factor A does not cause, but is "associated with" Outcome B. Cost is still high because of controls necessary.	1. Subjects form natural groups or else volunteer. 2. Control groups (no treatment) are possible.	Less than 10 percent
Anecdotal evidence for leadership Greenleaf's ideas about what constitutes a good leader (see Chapter 10)	No statistical proof is ever possible that a theory or idea about leadership is true. Cost is low.	1. No assignment to groups. 2. No control group. 3. There is never "proof" or any probability that Leadership Factor A causes Outcome B.	Roughly 90 percent of leadership literature is anecdotal.

Figure 1.4 Contrary to conventional wisdom, most available material about leadership is anecdotal.

Figure 1.4 shows that while is it possible to "prove" a theory about leadership by using *true experimental* methods, it is exceedingly difficult. For example, you can't ethically and randomly assign one group of test subjects to suffer for years under "toxic" leadership.

More often, using *quasi-experimental* methods, a theorist can fairly well establish a statistical basis for a theory about leadership using an alternate, vicarious variable for which data are available to stand in for the actual variable for which there are no data. For example, and as we will see in later chapters, Jim Collins uses financial and economic data suggesting that truly great leaders organize their groups in certain ways that make them financially successful.

The study I cite later on the Buffalo Police Department in Chapter 12 will suggest that officers suffered from psychological and physiological stressors, which resulted in sleeplessness and heart disease, but the investigator concluded that true leaders learned how to do deal with these stressors among their subordinates. The study is necessarily quasi-experimental because the investigator can't ethically subject a given police officer to stress (investigating car crashes or homicides) for 10 or 15 years. Instead, the officers self-selected (desk duty vs. patrol duty, for example) and formed natural groups with greater or lesser degrees of daily stress while working. From this, the investigator inferred (not proved) that stress is associated with cardiovascular disease, among other things. This is compelling (associative) evidence, but not perfect (causal) evidence that a hypothesis is true.

And honestly, quasi-experimental research on organizational behavior and leadership is about as good as it can get: It still is costly to operate a study of any size; the Buffalo study is eight years old already and far from completed.

Survey data can fit in this category and provide evidence about a hypothesis. It is much easier and less expensive to gather but suffers from the same experimental deficiencies as quasi-experimental studies: No control groups and no random assignment of subjects to groups are possible. But still, survey data—even with caveats—have some value.

Anecdotal evidence to support a proposition about leadership theory is pretty much everywhere: Airports are full of quick-reads about what the author thinks is true, but in fact, actually, these hypotheses are rarely, if ever, broken down into an experimental design; they have no control groups; they have no random subject assignment. On the contrary, they are merely good and captivating reading rather than a treatise on organizational behavior. Robert Greenleaf's work on servant leadership is much respected around the world (see Chapter 10), but his work is not experimental in any way. Robert C. Maxwell has had a bestselling book on the market titled *21 Irrefutable Laws of Leadership*, whose title says it all, but still, his conclusions are not experimental at all.

Which should a junior leader use: Data or stories?

Just because a compelling author uses nonexperimental methods doesn't make it bad. Its conclusions are simply not provable, that's all. And there is so little real experimentation going on in organizational behavior that we'd have very little to give to subordinates if we used only that.

Almost certainly, a blend of experimentation and anecdotal theory is best to provide to subordinates. We'd always like to have perfect experimental tests to actually prove that Leader Variable A and Outcome B, but that kind of conclusion is rarely possible and rarely ethical. Leader stories are emotionally compelling, and they captivate an audience or workers who are less interested in p values or correlation coefficients than the conclusions.

Fine, then. The best leaders will blend hard evidence with good stories.

In academic safety programs and in engineering, we have concentrated on management and devoted little time to leading people. We're more comfortable with hardware, Occupational Safety and Health Administration (OSHA), fluid dynamics, instrumentation, compliance issues, and training than leadership. In my view, that was fine when our graduates were 35 to 40 years old and already midway through their careers. That isn't the case anymore. Expectations for our youth are much higher now, and so are the consequences for bad decisions.

Academics have regularly used historically significant books and ideas to raise our collective sights for a better tomorrow. We probably all recall reading in college *In Search of Excellence: Lessons From America's Best Run Companies* by Tom Peters and Robert Waterman, which was originally released in 1982 but re-released in 2004. It was all about maintaining close customer–vendor relationships and the best leaders having a "bias for action," for example.

Not much later, in the late 1980s, corporate America woke up to the fact the Edwards Deming was living only a couple of miles from the White House. He was the same person who almost single-handedly transformed Japanese manufacturing into the envy of the world after he started there doing population counts at the end of World War II (WWII). His publication *Out of the Crisis* in 1982 and then again in 1986 and others (Deming, 2000) directly prompted Ford to build the Taurus; his work prompted General Motors (GM) to build the Saturn; and his work prompted Harley-Davidson to eventually emerge as the envy of motorcycle manufacturers around the world based on concepts like *kaizen* and statistical process control.

If you're interested in cars and things mechanical, I suggest you read Peter Reid's book, *Well Made in America*, published in 1991, for a wonderful exposé about Harley's transformation using the Deming processes.

Deming's work as a whole is again necessary but not sufficient to mold tomorrow's safety and engineering professionals.

I have visited the Honda plants and the Harley plants around the country. The Honda execs wear the same uniforms as line workers and the Harley guys wear, well, tattoos. Yet each worker can discuss x-bar charts and calculate upper and lower control limits for hours on end. Each understands continuous improvement and each can discuss the need for inspection at every phase of the manufacturing process and not just at the end of the line, like the places I have worked in industry.

But despite most college graduates having read these seminal texts, there is little or nothing in them about leader development or how junior leaders should act in difficult circumstances; it seemed something was still missing from our curriculum, something students and entry-level managers could implement themselves, even when upper management didn't seem to care. Something directed squarely at safety and engineering professionals has been missing about leadership.

Recently, it struck me: Maybe this leadership stuff, at its central core, shouldn't be really that hard to understand. For now, let's draw a simplified distinction to serve as a building block as we begin a discussion of leading subordinates.

The work of Peters and Waterman and Deming's work have transformed America's manufacturing processes and brought America's industrial might back to the forefront in world markets. There are solid reasons behind people wanting to buy a Buick in the Far East in preference to Japanese brands. But those seminal texts didn't help us understand leadership and leader development. They didn't discuss ethics and they certainly didn't discuss business protocol. Newer books, discussed in forthcoming chapters, do a much better job and bring us into the new millennium.

And then there is the U.S. military. They don't assemble products and they don't have manufacturing philosophies, per se. Yet they do "manufacture" some of the best leaders in the world, and they do "vend" their work products to American industry. Wendy's, Johnson & Johnson, Procter and Gamble, Goodrich, and Foot Locker have all had academy chief executive officers (CEOs), and in American government, Jimmy Carter and John McCain come to mind.

Fortunately, military leaders make the study of leadership and leader development reasonable and rational and we can learn a great deal from them. I borrow heavily from military leaders in this book as we search for what can work with our particular young people: Ours are not accountants, ours are not art majors. Safety professionals and engineers are involved in a business pursuit, exactly like the military, where people can die when wrong decisions are made. You'll see this phrase and comparison used here often.

A clear distinction would be good about now

Let me propose a simple definition of leader. Much of what we have learned about leaders and leader development has come from studying how the military organizes and trains leaders. Yes, I know the military is different from industry, but we often ignore their contribution to organizational research (see Figure 1.5).

And even though I am apparently standing alone here, I think it's about time that people in the safety and engineering professions take note of what the military has done in research on leadership and leader development. We are in pretty much the same business and use pretty much the same sort of motivated young people upon entry to our respective professions.

And while it might look like I am belaboring the point, I am surprised and saddened that safety as a profession has pretty much ignored that the military institutions in the United States have studied, researched, trained, and fostered leader development for over 200 years. Harvard Business School, among others, has arranged field trips to Gettysburg for tours led by military experts for years. Why? Their goal is to investigate leader decisions under the most challenging conditions imaginable—even conditions where people can, and at Gettysburg, they certainly did, die. Was Lee too bold and Longstreet too cautious? Was Chamberlain an authentic leader or just plain lucky at Little Round Top? Reading outside the classroom will begin to show us the strong link that safety and engineering have with the military's careful preparation of leaders in this country. *Killer Angels*, by Michael Shaara (1974), the Pulitzer winner from 1975, is still highly recommended by both civilian and military experts who understand what makes up the best leadership qualities. The book can shed light on the value of leaders under dire circumstances. The military, most engineers, and safety professionals share identical goals, it would seem to me. All three professions seek to preserve and protect the people, property, and efficacies of their respective organizations.

In the simplest terms:

A manager takes care of business. A leader takes care of people.

A manager does things right. A leader does the right things.

Figure 1.5 A good leader is a good manager by default.

The quote is from a well-known sociologist and former professor at West Point, Col. Tom Kolditz, speaking about what the nonmilitary world can learn by looking what military leaders are doing:

> …military leadership qualities are formed in a progressive and sequential series of carefully planned training, educational, and experiential events—far more time-consuming and expensive than similar training in industry or government. Secondly, military leaders tend to hold high levels of responsibility and authority at low levels of our organizations. Finally, and perhaps most importantly, military leadership is based on a concept of duty, service, and self-sacrifice; we take an oath to that effect. We view our obligations to followers as a moral responsibility, defining leadership as placing follower needs before those of the leader, and we teach this value priority to junior leaders.
>
> Soldiers in such circumstances must be led in ways that inspire, rather than require, trust and confidence. When followers have trust and confidence in a charismatic leader, they are transformed into willing, rather than merely compliant, agents. In the lingo of leadership theorists, such influence is termed transformational leadership, and it is the dominant style of military leaders.
>
> The best leadership—whether in peacetime or war—is borne as a conscientious obligation to serve. In many business environs it is difficult to inculcate a value set that makes leaders servants to their followers. In contrast, leaders who have operated in the crucibles common to military and other dangerous public service occupations tend to hold such values. Tie selflessness with the adaptive capacity, innovation, and flexibility demanded by dangerous contexts, and one can see the value of military leadership as a model for leaders in the private sector. (*Harvard Business Review* [*HBR*] blog, February 6, 2009)

Let me summarize to this point. So far, I am making the case that:

- Academic institutions have bypassed teaching leadership in safety and engineering curriculum, especially any that is very much research based.

- Career paths are evolving and graduates are challenged far more than ever. They are soon faced with high-risk, high-consequence jobs.
- We can learn from Peters and Waterman or Deming sorts of high-ideals examples, but we should also discuss how to enter a profession, ethics, office, and business protocol, and above all others, what a good leader does. Some of what we learn about what a good leader does can be learned by studying military examples.

Are there actual data to suggest the need to study leadership, ethics, and protocol among our future professionals, particularly the Millennials?

There are such data, yes. One such source of data is the anonymous annual exit survey we conduct as our Master of Science (MS) students graduate. We want to know which courses served them well and which ones did not; we inquire about instructor strengths and weaknesses. The exit survey is probably no different from similar surveys at other schools.

Along the way, it became apparent that the demographic had changed in academic programs. Figures 1.6 and 1.7, from my 2012 presentation (see Winn, Giles, and Heafy, 2012) at the Denver, Colorado, American Society of Safety Engineers (ASSE) Professional Development Conference (PDC), illustrate some important trends about Safety Management (SAFM) students.

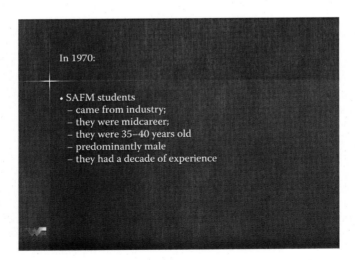

Figure 1.6 When most academic safety programs began, students were already midcareer.

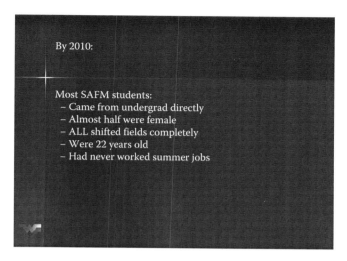

By 2010:

Most SAFM students:
 – Came from undergrad directly
 – Almost half were female
 – ALL shifted fields completely
 – Were 22 years old
 – Had never worked summer jobs

Figure 1.7 But by 2010, the demographics of academic safety students changed radically.

In an effort to find out what other demographic differences existed between the former and newer student cohorts, I led a small research group at WVU in the Statler College of Engineering and Mineral Resources, and we confirmed some suspicions we had about work and management experience, replicated over five years.

What we discovered over that the five-year period led us to tentatively conclude that our graduating MS-level students had little management experience or industry experience. Approximately half had no industry or management experience whatsoever.

My research team decided to broaden the survey by adding two more schools with engineering, history, chemistry, and other majors. We wanted to see if these results replicated about industry and management experience, but we also wanted to see if the students appreciated the need to know about ethics, business protocol, and leadership. The full results are available in our paper, *A Research-Based Curriculum in Leadership, Ethics and Protocol for Safety Management and Engineering Students,* which we presented to the Las Vegas PDC in June of 2013. A couple of sample figures follow.

Of the 53 respondents, 79 percent indicated that summer work was "sort of important" or "very important." Yet, there is an observed discrepancy between those working and those saying summer work is important. Only about half of our respondents reported working in the summer (see Figure 1.8).

Respondents in this same survey seemed to have good ethical judgment. When asks what a survey respondent would do in a situation where a fellow employee clearly acted unethically, (for example, taking

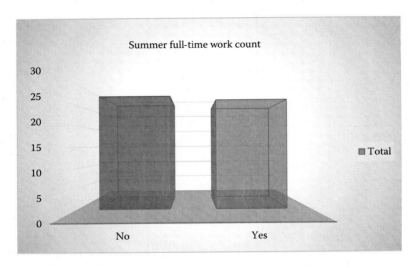

Figure 1.8 By 2012, students at three university academic safety and nonsafety programs had never worked a full-time job.

an expensive gift from a contractor they knew company would they knew later hire), fifty-three percent would either inform the boss or else let an employee outside the department deal with the ethical breach. Only 4 percent would take a passive route of informing an outside body and essentially walking away from the situation (see Figure 1.9).

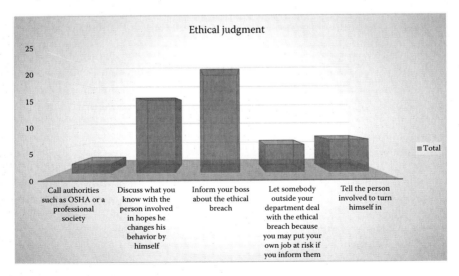

Figure 1.9 Students in the same survey had a good idea of what constitutes ethical judgment.

Finally, in this same study, we asked participants what they thought about the value of leadership training during their college career. An overwhelming 98 percent of participants—the highest response to any question in the survey—indicated that learning about leadership in the career preparation was "sort of important" or "very important in a person's career path" (see Figure 1.10).

Perhaps not surprisingly, an identical number of survey respondents (98 percent) indicated that the opportunity to practice what they might have learned about leadership outside the classroom was "sort of important" or "important" (see Figure 1.11).

Taken as a whole, the data suggest that the about half of the students we surveyed have missed opportunities where they could have learned about management and office protocol. The respondents seem to have a good inherent ethical judgment based on the survey, and they do appreciate the importance of leadership to advancing their careers.

Our survey also asked about reading and travel. Seventy percent of respondents indicated the importance of reading a daily newspaper as "important" or "very important." Further, we asked students in the survey about how much they value reading a weekly news magazine, and 66 percent responded that it was "very important" or "sort of important." For reading histories and biographies, students responded in the same fashion, with 72 percent valuing it as "very important" or "important" to one's career path.

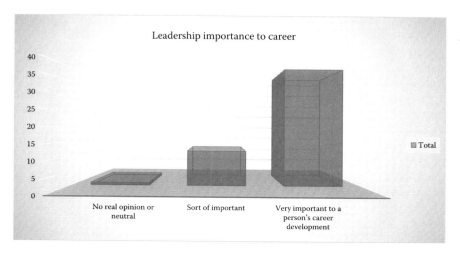

Figure 1.10 Even before they enter a profession, safety and engineering students understand the need to grasp fundamentals of leadership and its importance in their career.

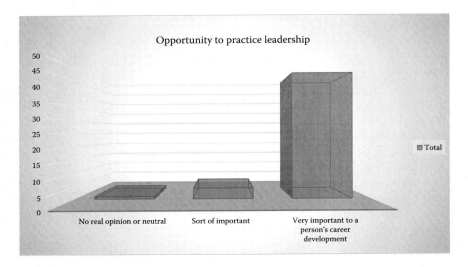

Figure 1.11 Students also think that it is important to be able to practice leadership before they embark on their new careers (*n* = 55).

But strikingly, the respondents in our survey stated that they only actually read a daily newspaper once or twice a week (38 percent), with weekly news magazine (26 percent) or histories and biographies (19 percent) even less. Clearly, there is a big discrepancy between how much the survey participants value reading for their career development and how much they actually do it.

The participants in our survey seemed to clearly understand that they were going to have to travel widely in the United States in these career paths. Seventy percent indicated that they had traveled out of state "many times," and 74 percent of these respondents thought that out-of-state travel was "sort of important" or "very important." But while only 19 percent of respondents had traveled out of the country "many times," 64 percent knew that out-of-country travel was "sort of important" or "very important" to their career paths. These students aptly predict in these four questions that travel, whether in-country or out-of-country, will figure prominently in their career, and while they don't travel a lot, they know it is important career-wise.

Unless we change our curricula, new safety and engineering grads will not have been exposed too much, if any, academic discussion of ethics or protocol. They will not know much about crisis and noncrisis leadership models. They will not have a research-based exposure about how to change corporate culture. They will not know the differences between training and experiential training, which the military so expertly applies. They probably won't know the difference between managing and leading.

In summary, I have predicated the need for this book on trying to make up for missed opportunities. Most students in our surveys do not work in industry, management, or summer jobs: These are missed basic opportunities to pick up experience with basic office rules and procedures.

Our survey participants seemed to understand the need to travel and read, but many didn't do either. In careers where future military officers and safety professionals are going to travel widely, survey participants indicated that they have traveled widely in the United States (70 percent), but not internationally (19 percent), even though the vast majority suggested that travel was important (74 and 64 percent, respectively). We see this as a missed opportunity to learn about other cultures, about business, about economics, and politics.

Further, our 2012 survey participants understood that professionals must read widely, but at this stage of their career preparation, they did not. These young people knew they should read daily newspapers, weekly news magazines, and histories and biographies to stay current on global national and international politics and economics because their future work places are increasingly abroad. Only about one third of survey respondents actually read newspapers, weekly news magazines, and histories regularly. This represents a missed opportunity to gain economic, cultural, and historical perspective.

Finally, survey participants told us that they wanted to learn about leadership models. In the pair of questions ranking at the highest affirmative response of the entire survey, our survey respondents indicated that learning about leadership during career preparation was "sort of important" or "very important in a person's career path." The same extraordinarily high proportion, 98 percent, suggested that practicing what they might have learned about leadership outside the classroom was "sort of important" or "important" in their particular career paths. Our survey results told us that graduating students know that learning about leadership and its practice is valuable to these respective career paths, including safety, engineering, social, and behavioral sciences, among others.

It's time for our young professionals to go to the next level: From managing to leading

Over the years, I have taught management principles and industrial hygiene instrumentation. I have taught construction management, safety training, and even freshman engineering. I am the luckiest person I know because I am surrounded by young people eager to learn about addressing social problems including saving lives and preserving resources. I still teach the basics with a focus on management, and yes, I still hold the occasional knife-making workshop on weekends, just for the record. But the last few years of talking to recent graduates and industry vice presidents have suggested that I concentrate on teaching another set of skills to our students—a new set of skills that will more quickly advance our students, save even more lives and resources, and get their office moved next to the CEO. These are, of course, my ultimate goals. In this book, I want to talk about leadership and leader development for young people entering safety management and engineering. These are the skills I think will set apart our students and offer them a chance to rapidly move up the ladder more than just another course in regulatory compliance. My research suggests that it isn't enough to just teach management any more.

The past was about managing. The future is about leaders, leading, and developing leaders. Why?

A manager takes care of his or her business. A leader takes care of his or her people. But academic programs spend very little time, if any, on motivating or disciplining subordinates, for example, much less teaching them to deal with "toxic" leadership or events such as an unexpected fatality. We need leaders for that.

A manager does things right. A leader does the right things. Yet academic safety and engineering programs are silent on seizing leader development opportunities in everyday training or considering "business continuity" in a way they can explain it to the CEO, for example. These are not part of the compliance model on which safety and engineering programs would normally limit their focus. Again, leaders are needed to see beyond just today.

In a global economy, in a socially connected world, in a culturally diverse workforce, in a technologically driven world, those old models simply do not prepare students well enough anymore.

The reason and need for a new approach to safety management have been over a decade in the making. Somewhere in the late 1990s, it became evident that our student demographic had changed. In a paper we delivered to the ASSE recently, my own research team noted the impact of the changed demographic (Winn, Giles, and Heafey, 2012).

Forty years ago, SAFM students at WVU were most often employed in vibrant Appalachian-region industries such as steel making, the chemical industry, or in manufacturing. The regulatory aspects of the safety movement being brand new, these midcareer students were encouraged by their company's upper management to seek a graduate degree so that the company could fill compliance and programmatic roles required by the new Occupational Safety and Health (OSH) Act and by state governments with local programs. At the time, SAFM graduate students were mostly male, in their mid- to late thirties, and they had a decade or more of industry experience after finishing undergraduate degrees, usually in physical education.

By the mid-1990s, most of general industry, followed by construction companies, had fairly mature safety and health programs. Many companies added an environmental responsibility to the safety professional's job, and many industry programs became the EH&S or SH&E programs that we know today. Following suit, our graduate students changed, too. They did not spend time in industry before coming back to school. In fact, by the late 1990s, these new students had much more in common with undergraduate students, preferring day classes to evening, for example. With little experience in industry or safety, these new students would graduate in May and right away apply for Fall admission to SAFM. They came from undergraduate majors in agriculture, animal science, business, parks and recreation, wood science, exercise science, and civil engineering, areas in which jobs in Appalachia were slowing.

For faculty in many academic safety departments in the 1990s and the early 2000s, students coming directly to graduate school with technical undergraduate degrees had positive consequences. The new 22-to-24-year-old SAFM student was comfortable with campus social life, social networking, and working in groups. The hectic pace and family life of our earlier midcareer students with 50–60-hour work weeks were unknown to this new breed of students. They had never worked much, we discovered.

Coming directly from an undergraduate school, it became apparent that these younger graduate students did not know important things about industry that would impact their EH&S skills. They did not know about work measurement, human resources, time and motion studies, incentive pay, or issues with unions. They did not know what a drill press was, or why respiratory protection might be important in welding operations. Since this knowledge may be learned in class but is usually acquired on the job, lacking it at the outset of a new job may have served to limit career advancement or even prevent students from obtaining some safety and health jobs in the first place. To offset some of these key skill and interpersonal shortcomings and to give industry a chance to "interview" interns in great depth usually over a summer, many academic safety programs established internship programs where the graduate student works in an industry of his or her choosing. Almost right away, this field experience was recognized as a 12-week job interview and good for both students and companies alike. By the year 2000, all other academic safety programs in our region also required internships (Winn, Giles, and Heafey, 2012).

The importance of the internship experience to pre-professionalize safety students cannot be overstated, but I judge that it is not enough. Consisting of a summer semester in most cases, the internship is far too brief; it is not consistently applied across industries and supervisors and does not speak to differences between managers and leaders but for students with little or no work experience, the internship is a good start.

Is understanding the difference between managing and leading really all that important? In a word, yes.

By most accounts, a manager applies the science and behavioral requirements of his job to the degree specified. That is, a safety professional or project engineer knows the technical requirements or regulatory applications needed to comply with a fall protection standard, for example. A safety professional or project engineer knows the science of air monitoring to create and operate a respiratory protection program. In my mind, the country's top safety programs have successfully created legions of what I call (somewhat disparagingly, I admit) "safety cops," who are the central players in preserving and protecting the people, property, and business efficiencies needed for a company to survive and compete (this was formerly the mission of the academic safety program at WVU, and it is shared in similar wording by other programs across the country).

But even an effective safety professional is something of a bureaucrat in the sense that having met his or her obligation for technical competence and meeting the letter of a given regulation (whether it's in a particular section of the code of federal regulations or knowing whether to specify an S or a W steel I-beam for a project engineer), the manager stops having met the requirement. What happens when the safety professional meets what Dr. Tom Kolditz calls *in extremis* situations (see Kolditz, 2007), where leaders are called to do things that never appear in some textbook?

This was the situation described by Kolditz, where a hospital director responsible for patient safety and health during Hurricane Katrina decided to commandeer buses and oxygen bottles sufficient to save patients who would have certainly perished without his leadership in taking charge himself. A manager would have called Federal Emergency Management System (FEMA) and waited, because *a manager takes care of his or her business.* The manager would have clearly and correctly met his obligation by calling FEMA. *A leader takes care of his people.*

In New Orleans, the hospital administrator—the real leader—couldn't leave people in somebody else's charge and hope for the best. He acted proactively. He "borrowed" school busses to move people out. He saved lives by leading and motivating change. In a crisis (which we'll see in Chapter 10 varies on probability and degree of consequences), we want leadership, not mere management.

In a crisis, would you rather work for a safety manager or a safety leader?

When it comes to people, and then when it comes to *crisis situations,* managers don't have a dog in the hunt; they just don't understand what comes next. Leaders do. That's the difference between the two, and it forms the foundation for the remainder of this book. *Leaders take personal responsibility for the ultimate welfare of their patients, their employees, or as we will soon see, their soldiers.* The task and methods are strikingly the same.

Col. Bernie Banks, now the chair of the Behavioral Sciences and Leadership Department at West Point, notes clearly the need for not just managers who can get things done in an administrative setting but also leaders who can operate when the lines are fuzzy and when people's lives are at stake—often the case with safety professionals and engineers.

"Today's organizations operate in what the Army War College defines as a *VUCA environment. Volatility, uncertainty, complexity, and ambiguity* are constant realities in the 21st century. The military seeks to prepare for the challenges it will inevitably face by crafting realistic training scenarios and routinely integrating such activities into its ongoing operations. The goal is not to teach them what to think, but to enhance their ability to think critically and creatively about the myriad of contingencies posed by a fluid environment—in essence to teach them how to think," says Banks in the October 28, 2010, issue of the *HBR*.

In the remaining chapters, I explore leadership styles and models available to learn from, drawing from industrial and military thinking. But as I have learned from talking to some of the very best minds in the country about leadership and leader development, we have to set the stage first. First, we have to discuss what it means to be a professional safety professional or engineer, and to get even that far, we have to talk about getting a dinner invitation (wait for it). Then I discuss some discoveries about personal growth that I see as necessary for any student of leadership, but particularly my Appalachian students.

Then we move to self-awareness and value systems; next, we explore applications of ethics in safety and engineering using canons of ASEE and the National Society of Professional Engineers (NSPE) as important models, following up with leadership theory in both extremis and non-extremis situations.

Finally, we will review some tenets of professional conduct and protocol using models from the military, a dichotomy I'll refer to again and again when I ask "why study military leadership and protocol?"

Because they've been in the business of developing expert leaders since 1802, that's why. See the January 2014 issue of *Professional Safety* for a current view for safety professionals and engineers to study military organizational research (Winn and Banks, 2014).

Our graduate programs have been successful in placing young men and women in the toughest and most challenging industries in the United States and, more recently, the world. Salaries have climbed rapidly. The National Institute for Occupational Safety and Health says there will be two safety jobs for every graduate for a decade to come (NIOSH, 2011).

Now, it's time to amp up the preparation of our grads as we bring leadership and leader development to the individual first and, assuming that preparation is sufficient, allowing those graduates to develop their own leaders in their respective industries. As part of a strategic plan to improve graduate education and push leader development to the forefront, the SAFM Visiting Committee, composed of recognized graduates—business leaders—has even endorsed the change in our very mission, which now reads, "to create leaders who protect and preserve the people, property and efficacy of a business or corporation."

The days of being "safety cops" worked well, but those days are past. Let's think about the future. Same for engineers.

Sorry again about having to say "safety cop." I want to make it sound disparaging, I guess.

Do others recognize the need for change? Indeed, they do

Mrs. Frances Hesselbein is a wife, mom, and homemaker from southwest Pennsylvania, with Appalachian family roots going back to well before the American Revolution. Mrs. Hesselbein is the president and CEO of the Frances Hesselbein Leadership Institute (formerly the Peter F. Drucker Foundation for Nonprofit Management). She was awarded the Presidential Medal of Freedom, the United States' highest civilian honor, by President Clinton in 1998.

Mrs. Hesselbein knows a thing or two about leader development. That's also precisely why we are discussing her textbooks in my leadership class (Figure 1.12).

Figure 1.12 Frances Hesselbein visits with Dr. Winn at Virginia Tech in 2012.

At a recent conference at Virginia Tech, I caught up with Mrs. Hesselbein, who was disillusioned a bit about prospects for the future. She said during a presentation that she thought the current generation of college graduates expresses the "lowest level of trust and the highest level of cynicism in my own country, in my own lifetime." Employees distrust their leaders and are cynical about the system, she suggests. By itself, this sentence alone spoke volumes about the state of the art in preparing young people for a lifetime as a motivator of people, a planner, an organizer of activities, a harnesser of human energy.

Yet far from discouraged, she also suggested that the current generation of youth is the "generation that will save society [from cynicism]— you will shine a light in the darkness" by serving, by leading, and by instilling leadership by serving others first. And this is exactly where I am going with this book when she said at the same conference where I met her, "to serve is to live." For my safety professionals and young engineers, you might as well tattoo something important on the back of your hand, like the foregoing advice. It's that good.

Like most experienced academics, I think a lot of my graduates, and there is a lot to be proud of. All of our graduates have committed to a higher purpose before they walk across the graduation stage to shake hands with our dean and chair—saving lives and resources for the safety professionals and fulfilling human and social needs through math and science applications for the engineers. Grads are being employed in a broader range of industries than ever now, including aerospace and petrochemicals including NavAir, Chesapeake, and CNX. They are offered jobs in the highest profile construction companies in the world including Parsons, AECOM, and URS.

Our grads are offered positions in international companies, including International Hotels Group and Royal Caribbean Cruise Lines, Chevron, and Schlumberger. Our interns are serving in Dubai, Guam, and Qatar.

Regarding the protocol of being a professional, I've found that, sometimes, simply telling my students what to do often is sufficient for behavior change. Show them the standard and I am confident they will reach it if they are shown the best way.

Raise the standard higher and they will most often meet that, too. But a decade and a half into the new century, they still have to be educated about what behavior is proper and what is not (Figure 1.13).

In recent years, I see more and more students who come to my office about enrollment in our graduate program with a hat on backwards or without paper or pen to take notes on their academic plan. I fix these basic missteps of office protocol by simply talking about it, about "dressing for the job you want" and how to make a good first impression. Often, I hand out a simple "flying WV" portfolio in our very first meeting together and wonder why, by age 22, this behavior persists at all (Figure 1.14).

Figure 1.13 Not all schools pay attention to training their students in etiquette and protocol. At the Virginia Military Institute, they do.

The phone call from a head hunter about the new hire with a six-pack hanging on his belt at an industry conference still haunts me.

I have little patience with students who want to say that business etiquette is a matter of "personal preference" or that business protocols are relative. For God's sake, business protocol standards *are not relative.*

But if we professors don't tell them what is expected, or worse, they begin a new job without experience to show them particular standards of behavior and business protocol, then we'll find underneath a fresh-faced, well-intended young person who doesn't realize that it isn't appropriate to be late to a business meeting for any reason.

And then there is the spit cup...

Yep, it's gross. I am told that using a spit cup is a way to spit indoors and use tobacco at the same time; I've never tried, so I don't honestly know. In

Figure 1.14 Pre-vet student turned safety professional Kate O'Hara and her friends enjoy a spring afternoon. Kate is responsible for the safety and health of 320 people every day in a top-secret environment. Kate is a leader for the Millennial generation.

days gone by, students left cigarette butts in my office, but these have been replaced with the spit cup, or a small plastic water bottle which serves the same purpose as the cup.

I thought I was the only professor who had a small collection of spit cups and bottles in his office left over from advising appointments. I was wrong.

I shared the spit cup story with Col. Thomas Musteen, a professor of military history at West Point and a man I greatly respect because he is a hell of a marksman and an expert in the French and Indian War. He recently gave up his position at West Point and a follow-on tenure track position at an Ivy League School to lead his troops as the commander of the last armor company in Iraq. Yes, he's my hero because that kind of behavior shouts "leader" and not just "manager."

And as an assistant professor at West Point, Col. Musteen knows full well about spit cups. "Thanks for bringing this up," he said to me recently. "I thought I was the only professor with problems with my students bringing spit cups to class," says Musteen. "My one-man fight against spit cups has had somewhat disappointing results in the Army. However, you've made me feel better because my assumption was that the Army was the only subculture in America where educated adults could pretend to be professionals while spitting into bottles and Styrofoam cups during meetings. Now I know the problem is bigger than I thought. And that there are others in the fight with me."

I am proud to join the battle of the spit cup, sir.

Thank goodness that we professors are not alone in the battle to fight the spit cup or to improve business etiquette. But business etiquette in a general way is yet another reason to begin to teach our students to step up and think of themselves as more than safety professionals or project engineers.

Not knowing basic business etiquette means that when you hang a six-pack on your belt at a conference, you may as well just fire yourself. Ask my former colleague, Kevin S., who thought that particular behavior was OK at a small conference.

I can only hope that days like that are over, but the point about the importance of courtesy, decorum, and uniform business practices is important to grasp early in a career.

A graduate trained in basic leadership and etiquette skills cannot only improve his or her professional status in the eyes of his superiors, but also he or she will show his or her subordinates how to act as well. That training and model-setting will probably go home with workers. As part of leadership and leader development, we're just flat going to talk about business etiquette including how to write a basic memorandum. Why bother with the lowly memorandum? Because a memo is the most fundamental tool used in office communication; because a memorandum is a perfect mirror on a person's professional composition skill, and because nobody has shown young people how to do it in 14 years of education, that's why.

We have now set the stage and established the need for this book: to help young people entering the safety profession or engineering to understand leadership, ethics, and protocol before they enter a fast-changing workforce where the challenges—and rewards—are great.

And so, this is the final and maybe most compelling purpose of this book: to introduce some basic rules about business protocol and dress for people who missed opportunities along the way to gather it for themselves. That's why that chapter on business etiquette comes at the end of the book, and not now. So let's get started.

chapter two

Self-discovery comes first

Why don't we just jump to talking about leadership? Early in writing this book, that was my question, too. But in my own impatience, I side stepped some very big issues that have confronted experts also looking to create leaders of character. I guess I discovered that we need to "go slowly in order to make haste."

One of the biggest mistakes I made starting out was thinking training on the creation of leaders of character can be condensed into a book or a semester. In a series of interviews I did in leading up to begin writing, I travelled to the home turf of experts in leader development. At the Virginia Military Institute (VMI), to start, I interviewed Brigadier General (BG) Casey Brower, who is the former deputy superintendent for academics and dean of the faculty at VMI. BG Brower is also the former chair of West Point's Department of Behavioral Science and Leadership.

At my meeting, he was accompanied by Col. Tom Meriwether, a professor and well-known researcher in the VMI Department of Psychology. I asked what I needed to do to condense for a course or book, to distill what VMI does over four years in creating leaders of character who later will commission in the Army, Navy, and Air Force, and of course, the Marines. Looking back, I am gratified that these gentlemen did not snicker about my quest to reduce the essence of their multiyear task to single course and later, a book.

Maybe they did snicker after I left. I was pretty green in the process of understanding the path toward authentic leadership.

Col. Meriwether noted that the VMI program, began in 1839 based on a strong European system that valued leader preparation, although much of that system was based on birthright, not merit so creation of citizen soldiers and engineers became our goal. But as BG Brower noted, "both West Point and VMI have four years of embedded leader experiences, but the 'polar star' is [that we create] educated leaders of character." It started to become apparent that four years of experience would be difficult or impossible to recreate in leadership at our university. Getting to that level takes a lot of time and baby steps first.

When I met with these gentlemen, I noted to both Brower and Meriwether that we both trained youth in similar missions: that each young leader had responsibility for subordinates, each sometimes worked in the crisis mode, and each had budgetary responsibilities, too. Each had obligations to protect material and property resources and to support the efficacies of the company's business in a general way (protecting company image, for example).

Brower and Meriwether agreed with me, but Meriwether then spoke up, saying, "but you can't jump straight to leadership before individuals explore self-awareness." That is, authentic leaders *must* discover truths about how they feel about people subordinate to them and that young people at a military institution or not need to do this assessment pretty much for themselves. Self-discovery, they agreed, was an essential building block on the way to becoming a leader of character.

Finally, I thought. This was the place to begin.

Brower was more direct. The West Point model, he said, was the "gold-plated" version of leader development because the coursework comes just at the right time—before the cadet begins to practice the techniques as a sophomore training a single student in a semester and later as a junior when the student is responsible for up to 10 younger students. Long before leaders actually lead, and whether it's at a service academy or in industry, the students will examine human relations, motivation, what it means to be a "professional," and then values and ethics. Once these are under their belts, he suggested, it is time to present models of leadership. The early phases of leader development are "transformational" in the sense that the individual becomes aware of his or her own biases, strengths, and core values.

In my VMI interview, Brower and Meriwether agreed that self-awareness is the place to start, whether at WVU, VMI or even a small department in a wood products factory. Future leaders must know about their own primary motivations before they begin to address needs of their subordinates.

Public educational institutions—particularly higher education—don't have a fabric of 24/7 leader development, sorry to say. Most of the country's safety programs at the undergraduate and master's degree level and certainly the engineering and engineering-technology programs are accredited by the Accrediting Board for Engineering and Technology, now known simply as ABET. ABET which governs engineering curricula, or its branch accrediting agent, the Applied Science Accreditation Commission (ASAC), which governs safety-related curricula. ABET and ASAC curricula are tightly structured and required courses are specified. And while the notion of ethics is given some consideration by ABET, there is no requirement that the curriculum offer anything about self-awareness or values, much less leadership. We agreed at VMI that the best I could hope for, and far short of distilling what they do in 47 months at VMI, is draft a model for development over time. Same for industry: Leader development takes time and needs to be done in stepwise order.

On a later interview trip to West Point, I had the good fortune to run into the current chair for Behavior Science and Leadership Department (BSL) at West Point, Col. Bernie Banks, who helped me clarify further that leadership training is difficult to distill and must be done stepwise. It must begin with an examination of what the leader stands for—"what they need to *become*—and not what they *can do*." This advice from Col. Banks mimics that of Brower and Meriwether—start with self-awareness, then move to an examination of values, and then to ethics before engaging top gear with discussions of leadership.

Is there a similar model for slow and stepwise development of a supportive culture out there? Is there a parallel from the West Point/VMI "47-month model" to industry? I think there is such a model available for reference. It's Total Quality.

Recall that when Total Quality programs were first discussed, Ed Deming's advice was always to start with top management for a firm commitment to change the entire way a company did business, affirming allegiances to collecting and using data over mere good intentions, breaking down barriers between departments, and so forth.

The clear message from Total Quality proponents was this: Don't expect miracles overnight. Start by convincing upper management to fundamentally change and don't try to cherry pick the easy aspects of a total quality program. It's kind of "all or nothing." That parallel is almost a perfect analog for leader development.

Don't expect miracles overnight.

But plan on success eventually. Get on with it. Know what you stand for, and what you won't stand for.

At the university we don't have four years to prepare what Col. Banks calls the "ecosystem of leader development" (the full four-year experience for cadets at West Point), but a given company can still invest in a change similar to total quality and using similar principles. I call my model system "Value-Based Leadership for Safety Professionals and Engineers." Your company could call it something else, but under this particular title, it means that we don't just start in top gear discussing leadership models, even though some of them are strikingly useful, particularly under extreme situations.

Instead of starting at full speed, my research—and West Point's and VMI's—strongly confirms that leader development must be based on the model that begins with becoming a professional, moving through to self-awareness, exploring values, and then—and only then—how a leader acts. In the best of all worlds, every employee adopts the company's core values; every employee mentors less experienced employees in values; every employee studies ethics as related to the particular industry; finally, every employee begins to think of himself or herself as a leader; every employee's goal is to become an authentic leader of character.

I realized that the choice to become a leader is, first, an act of self-awareness, and it's also a person's first leadership decision.

I have adapted the graphic in Figure 2.1 from a thorough distillation of my interviews and research about leadership and leader development. It should help explain at a glance how an organization or learner can adopt the model I propose in this book. At the same time as learning about entry into a profession, the learner begins with self-discovery and, upon mastering it, moves to an exploration of values and values-congruent decision making. Moving forward again, the learner investigates how to make ethics-based decisions and actions before moving again into the realm of some subpractices that are sure to baffle safety professional and engineers if they do not study them ahead of time: avoiding toxic leadership, handling difficult employees, and managing the death event. Stressors to company morale come next, followed by a study of office and business protocol on the way to authentic leadership. If there is a single point to be made here, it is this: One skill set at lower levels *must be built* before attempting the next, more sophisticated level.

Authentic leadership

Fine tuning skills: understanding business challenges and office protocol

How experienced leaders handle stress and tests of organizational morale

How experienced leaders handle special challenges: difficult employees; the death event

Avoiding toxic leadership

Noncrisis and crisis-type leadership models

Ethical considerations upon safety or engineering decisions

Establishing personal and organizational core values

Exploring self-awareness and values congruency

Transition from student to safety professional or engineering professional

Figure 2.1 The pyramid for learning authentic leadership: the simple model.

Dr. Winn's 20 maxims about professional life: A first step in self-awareness

My Merriam-Webster dictionary defines *maxim* in a useful way: "a general truth or a fundamental principle or a rule of conduct." The maxims about entering a challenging profession have accumulated from my own experience—things I learned that have shaped my life and level of self-awareness. At worst, these should help the reader avoid costly, or worse, embarrassing mistakes in the first year or so of office life.

You can pretty well bet that I learned each of these the hard way in my years as I worked twice as an industrial engineer, as a lobbyist once, and even as I operated a small racing business for a while, where I watched some magnificently effective leaders first hand. Here is some of what I learned on the way to self-awareness of how a professional acts among peers. If you're paying attention, by the time you get to my age, you'll have 20 maxims of your own.

Maxim 1: Never complain and never explain. I'm sure I stole this maxim from someone years back, but I just don't recall who it was. I do know my students have heard me use it in class many times. It means simply that when you make a mistake, admit it quickly and completely but don't overexplain it. If you ignore your error or make too much out of it, you become a whiner, and that defeats the purpose of full disclosure. Trust me, leaders are not *whiners* and neither are whiners *leaders.*

Maxim 2: Don't let them see you sweat. Here, I only mean that you must be ready at all times for difficult situations. For example, in some companies, you must be prepared to discuss the previous day's progress when your boss calls upon you. In years past, when I was an industrial engineer at a helmet manufacturer, my operations meetings came at 8:00 sharp every day. Everyone, including production people, research and development, shipping, and quality control, all stood around the walls (no sitting allowed) in our conference room as the operations manager, an old Navy captain, grilled each person, saying, "Why is my life better today than yesterday with you on my payroll?" Ouch! After one day, I knew I had better be prepared to answer and even prepare my comments in my office for a few minutes. "Navy Jack" was tough, and I learned quickly to be ready to answer ahead of time. That way, I never let him see me sweat.

Maxim 3: Be prepared to achieve. Come to your boss's office ready to be successful: You're dressed as industry professionals do; you have paper to take notes; you have a company logo pad folio to look sharp.

You have a couple of questions ready. And when the meeting is over, you follow up with a quick memo to your file recapping the recent meeting; What message does it send when you're dressed badly, you don't have paper, you forget what happened at the last meeting? It says, "Please skip me when it comes time for promotion," that's what.

Maxim 4: Learn to say "yes, sir" and "no, sir." The single fastest impression you will ever make is the first time somebody hears you say "yes, sir." You will make a strong and positive statement about respecting yourself and respecting your superior simply by saying it. It might take some practice—you might have to force it out at first—but it will change your life. It also sends a message to your subordinates that you respect the chain of command and that it's something they should do, too.

Maxim 5: Be an achiever. I do not mean "be an annoying, cloying yes-man." I mean this: Be early to work every day; be early to every meeting; call in ahead of any meeting you have to miss and never afterward; never overstay your office visit; offer to shake hands first; offer your seat to a lady; hold the door for anybody entering the room. I have to say that I learned this from a friend who ended up as a National Association for Stock Car Auto Racing (NASCAR) executive and who, in his younger days, used to hold the best run rider and driver prerace meetings anywhere. Roger did not tolerate nonachievers and asked them to leave if they came late. He would not restart a discussion if latecomers arrived. If meetings were indoors, he closed the door at the appointed meeting time to purposely embarrass the late arrivers. It sends a message that doing things wrong is not ok. No break to Roger.

Roger detested non-achievers, and so do I. So should you.

Maxim 6: Be the "morning guy." Most work at an office gets accomplished in the morning. If you must leave for a doctor's appointment, for example, schedule it for the afternoon. There is just something implicitly less painful in going home early in the afternoon compared to the morning. Fewer people make notice of errands that need to be completed in the afternoon. However, people do notice morning absences, even excused ones. It just seems to be something about human nature, that's all.

Maxim 7: Do not be the "morning guy" with a travelling coffee cup. I have worked with people who have a regular route through an office, visiting and drinking a bottomless cup of coffee. A good friend here at WVU used to call them "the department's entertainment committee," wasting time with small talk and weekend's football scores, golf games, and other useless stuff. A little bit of that is required to qualify you as a normal social human being—I understand that. Just don't overdo it.

Maxim 8: Don't nominate yourself for awards. Larger companies and member associations will offer programs of awards, and recognition of achievement is surely a good thing. Although strange to me, some companies ask people to nominate themselves, rendering the award completely meaningless. Academic institutions, including mine, invite self-nominations for teaching awards, and so I plan to never have one.

I have faculty and industry acquaintances who have created pages and pages of material on themselves. I have seen four-inch binders produced in support of a person's own teaching or research award self-nomination. But in my world, an authentic leader nominates somebody worthy, never herself or himself. In fact, if one of my students or readers of this book—and I'm only being partly facetious—ever self-nominates, I will find you and burn your four-inch binder in the parking lot. Count on it.

Maxim 9: The importance of first names. I've done this for years, but a fellow I met at a West Virginia wood products company named Faron C. did it better than I did. Faron said even though he was new at the company, he planned to learn every employee's, all 368, names within six weeks. I was impressed, but when he also said he was also going to learn people's spouse's name and kids' names, I was bowled over.

People like to be called by their first name, and if you have to use some memory trick to do it, there are books for that. Faron didn't need a book—he just worked on it every day. As we walked from one part of his Moorefield, West Virginia, plant to the other side, Faron passed six or eight people and called every first name correctly.

I wondered if he was going to try to learn the kids' soccer team names, too. I wouldn't have been surprised.

A funny anecdote: There is a high-profile professor here in my college of engineering who has called me "sir" for two decades. (He thinks it's my first name, I suppose.) It has become a standing joke among even my students and even my academic colleagues and he obviously doesn't get it because the joke is on him. He can't remember my first name and he is embarrassed, but not embarrassed enough to try and learn it.

Maxim 10: Humility. You can do it. My buddy Dave Miller related a story he read about a custodian named Mr. Crawford working at the Air

Force Academy in Colorado in the late 1970s. He'd been there for years as a retirement job. An Academy student reading about WWII realized that the Medal of Honor winner from his textbook had the same name, Mr. Crawford, as their janitor; indeed, it was him. "That was one day in my life and it was a long time ago" was all he said. Rather than call attention to himself, he called no attention at all, and while he wasn't hiding his position as a Medal of Honor winner, he didn't think it was more important than doing a good job for his boss. That's authentic humility.

In Jim Collins' book *Good to Great*, he says that David Packard of Hewlett-Packard fame made sure his headstone would only say "farmer."

Maxim 11: Let your employees do meaningful work and let them innovate on their own. The hardest I have ever worked has been on projects that were meaningful to me (that is, matching up motivation with expertise or allowing a subordinate to develop a completely new expertise). When I was in the nonprofit sector, I had a boss who let me publish a scientific paper now and then because he knew I'd dig into a new area that was critical for the company.

Remember that someday, you'll have subordinates, too. If your employees see that their name is linked to a successful project that they initiated, they'll work hard on it.

As simple as recognition and meaningfulness of the work actually are, they go a long way toward maintaining employee interest when things are slow or more mundane. And the cost is zero.

Maxim 12: Be the "good mule." My grandfather farmed corn and wheat in Missouri, and he did it with two teams of horses but only one team of mules. My mom says that when conditions got difficult or the very hardest work needed to be done, my grandfather would pass by the horses and go straight to Ballie and Bess, his mules. They never refused to work and they always took on more than their share. Like the "good mule," if you never refuse a job, and if you always take on more than your fair share, your boss will take notice. After a while, you'll end up with more of the jobs you want to do and fewer of those you have to do.

Maxim 13: Authentic leaders pay forward. When I finished my doctorate at Ohio State University (OSU), I was fortunate enough to hear Woody Hayes, OSU's famous football coach, talk to the graduates. Besides being the best known coach in Ohio State history, Hayes was also an extraordinarily talented military historian. He never let that outshine his humility as a coach. The advice he gave us on graduation day has stuck with me for years now. He said that we can't pay back our family, our pastor, or Scout Master for their insights and help getting our lives on track. We can't possibly repay them. Instead, Hayes said, authentic leaders need to help the *next* generation with just about anything that advances their aspirations, whether it's career advice, the occasional financial help, or sometimes just time sitting in your office. This just makes so much sense.

Don't be afraid to take time out of your day to let people vent or confide. I have a box of tissues in my lower desk drawer for that reason.

Maxim 14: You are measured by how you treat the least among you. This is very important advice that I have borrowed from Randy Fullhart at Virginia Tech, and it stops me cold every time I come back to it. Do you know the first names of the janitors, the information technology (IT) support people, and FedEx guy? Or do you just know the vice presidents and department directors in your office?

Do you stop for a minute in the parking lot and just chat with the security cop? What about the fellow on the bus that looks like he could use a pep talk?

Once you see this maxim put into action, you'll know which leader you'll want to emulate and who is just a leader in name only.

Maxim 15: Embrace the suck. Even when the business climate takes a downturn, or when you have to be in the field at 5:00 a.m. for a training demo, do your job with as much genuine enthusiasm as you can muster.

In Tom Kolditz's book *In Extremis' Leadership* (2007), he says this about sharing difficult conditions to push boundaries:

> In situations when conditions are difficult or miserable but not necessarily dangerous, it's essential to motivate even the most dedicated individuals. Whether it's heat, rain, cold, filth, fatigue, deal with the misery intelligently and with some positive energy—in other words, "embrace the suck."
>
> If conditions are bad enough to be life-threatening, do not hide that fact. Use the motivating qualities of mortality salience [knowing that people can be injured or even die on this job] to push others beyond their perceived limits.

Maxim 16: Make yourself dispensable, not indispensible. Another way of saying this is what Douglas McGregor, the well-known organizational behavior expert, said: "surround yourself with the smartest people you can." He means, and so do I, that the work goes easier and faster when you can quietly back away and the job gets done.

Of course, that's not how some leaders act: They want people to think that the company itself will fail if they leave. They want subordinates to bask in their reflected glory.

A better model is the opposite, I think: Arrange conditions of the work place so the company goes on and the tasks get finished even if you get promoted or move on. If that happens, you haven't left the department or company in the lurch because you don't have an inflated concept of your own worth. You'll know you've done your leadership job well.

Maxim 17: If you watch the crash, you're going to be in the crash. I learned this truth at the racetrack during a 20-year amateur road racing career where I didn't win very often, but then, I also didn't crash much either. I realized that when somebody crashes in your immediate line of vision, it is very compelling to look at it, like the TV at the airport or the barbershop, except a racing crash happens sometimes at 120 or 130 miles per hour. And when you watch the crash, your body or your bike tends to follow where your eyes and your attention are fixed. Of course, a "crash" can be a business disaster or even a serious injury at your plant.

In racing, you must force yourself to look *past* the crash because the road and its confusion are still coming up on you fast—it's the same for a business setback. It's OK to take notice of the crash or business setback, but not so much that it pulls you into it; keep a safe distance and don't overreact.

If you can keep some psychological or physical distance from a setback, you'll be one of the few able to make important and timely decisions. This will take some practice, but it's worth the effort.

Maxim 18: Discover your own passion and enjoy it. Learn what your *personal passion* is and be comfortable in it. I don't mean being passionate about watching college football or, worse, golf. Everybody already does that, and it's a vicarious passion. And I do not mean being passionate and droning on about your job.

I mean developing a passion for yourself, a really intense hobby, no matter how obscure—maybe about learning to repair stained glass for churches in your region, or about helping you restore artifacts in your county's historical society, for example. I have friends who do each of these.

When you discover your passion, you will own it, and almost immediately, you'll find people that are drawn to your conversation. Maybe it's how the introduction of nonnative trout species may be related to genetically modified foodstuffs. You become *interesting*, and this one obscure passion takes the dinner conversation to international politics, and then to environmentalism and international agronomy. That conversation, in turn, may cement a business transaction because kindred spirits have found each other. Be proud of your own passion no matter how obscure.

I have a friend who, to the outside world, is a somewhat introverted and rather uninteresting guy until you know that his passion is stationary engines, those one-cylinder greasy giants of the oil fields of a hundred years ago. He taught me about the value of this kind of low-technology engineering and why it could be useful in third-world or developing nations even today, which in turn could be an essay right out of the *WSJ*.

Sherry, a 30-year friend of mine, started a small program called "Speak Through the Horse," which involves at-risk junior high and high school girls who share her passion for horses. She exudes the passion and the girls drink it in. With work and time, the trust between girl and horse builds a strong sense of self-worth and confidence. Character emerges in these young girls where it was hiding before. Annie, as an example, went from being a poor middle school student to one who was rewarded with "horse time" for improved study habits. Annie is finishing dental school now. Annie's dad credits her passion for horses and nothing else.

Maxim 19: Your mom was probably right: trust your instincts. In my many decades of looking for truth and good in people, I have been way too much a scientist. I've looked for data and mathematical algorithms and I've examined variables much more than I should have. Only in the most recent five or so years have I started to relax and think like my mom. She said a very long time ago about judging people's caliber: "Your parents and your pastor and your community raised you right, so trust what you know." If it feels wrong, it probably is. You don't need some needs analysis to know that.

Your gut feeling about people and their motivation is right most of the time. I should have listened to my mom on that one a long time ago.

Maxim 20: Create and update your Personal Professional Development Plan (PPDP). It isn't as important to just have a PPDP as it is to update it periodically. You'll know it's there and you'll know it is calling you to get busy with your life. Let your PPDP be a roadmap to keep changing and updating your skill set and networking contact list, and so forth. These are things a professional does, but without a roadmap, you just don't know when you get there. I've included a sample PPDP in the next section here as a draft. In class, I use these early in the semester, and again at the end of the semester. You can easily tell the people who are paying attention to planning for their future because their PPDP changes radically even over one semester. The PPDP is one way to make your future plans real.

The goal of the PPDP project is to formalize what your professional goals are and how you plan to achieve them. It is necessarily a "work in progress" because it is always changing. If you keep track of, and update, your professional goals and timetable, you have a target. My students redo this same plan at the end of the semester, but you can update it on your own once you're out in the field. Remember, if you don't have a target and a method, then *any road will get you somewhere.*

If you ask somebody whose judgment you respect, they will probably tell you that they have such a development plan for themselves, maybe not as formal as I propose here, but they do check off milestones occasionally and modify the plan as life intervenes, sometimes good, sometimes bad.

A reflective plan gives structure to a young professional. As I said, without a roadmap, any road will get your somewhere.

Dr. Winn's PPDP: A reflective plan

Answer these questions in short declarative sentences or a very brief paragraph:

1. What do you feel is your life's central purpose (for example, having a career that makes a difference in peoples' lives)?
2. What are your short-, medium-, and long-range goals for achieving your purpose above?
3. What do you feel is your life's real passion (for example, starting an adult reading program)?
4. What are your short-, medium-, and long-range goals for achieving your passion?
5. What career field do you see yourself most passionate about and why?
6. What are five methods of professional growth that will help you achieve in the career field you listed in #5?
7. What are three personal habits you hope to develop to achieve?
8. How can you foster the growth of each habit in #7?

9. What are three habits that do not contribute to your personal growth and development?
10. How do you plan to reduce or eliminate these less desirable traits in #9?
11. Name five core values that you use to guide your personal professional development.
12. Without using references, define what you think is meant by "leader of character."
13. Describe a landmark event that forever changed your life and why.

A PPDP isn't going to change your life. It's just a roadmap, always evolving and always being updated. You just haul it out every couple of months and see how you've changed. You'll smile at what you'll find, I guarantee.

In the next chapter we'll see that part of becoming a professional is to become interesting to others—to encourage them to join the conversation. Being self-aware, writing and updating a reflective plan, and reading extensively are three ways to start on the path to becoming a leader. The maxims you just read in this chapter are nothing sacred, only pieces of truth that have made me who I am today. You'll have your own to share and I encourage you to keep track of them. The reading suggestions in the next chapter are mine, too, but you should make your own list and update it periodically. Remember, knowing who you are and what motivates you is the first step toward becoming a leader.

chapter three

Further becoming a professional
It takes effort outside the classroom

What does it mean to be a professional?

Historically, the term *professional* comes from the root word "to profess" or "to promise," which suggests professing religious vows or promising to retain a religious faith. Indeed, the word "professor" has the same roots from when higher learning was limited to religious institutions. Now, the word *professional* still suggests faith, but not religious faith as much as promising to keep true to principles in an occupation.

For centuries, there were only three professions: the clergy, medical doctors, and lawyers. Teaching became a profession somewhere in the middle ages, and thus, there were four recognized professions for another three centuries. Later came nursing and pharmacy and, more recently, psychology, engineering, and social work. Even the military, whether officer or enlisted, has persons engaged in "the profession of arms" and meets the following definition.

In all cases, the term *professional* means an occupation (1) that you plan to pursue for years if not decades, (2) for which you are paid a regular salary, (3) requiring a high level of technical preparation, (4) requiring a high level of skill, and (5) implying a high level of public scrutiny and public trust.

Author R.W. Perks (1993) is often cited as establishing milestones with which a profession can be identified. I have modified the format a bit. A profession becomes a full-time occupation as opposed to part-time.

- A training school is established to teach specialized professional skills.
- Later, a university school is established for that profession.
- Accreditations follow at the university level (ABET, for example, if engineering) for that particular profession.
- A national association is established (ASEE or NSPE, for example).

- Later still, codes of professional ethics or behavior are issued for the profession.
- State licensing laws are established.
- Professional competencies are established (CSP or PE, for example).
- At some later date, professional competencies are recognized as job prerequisites.

Both safety professionals and the engineering profession have gone through these steps in the last hundred years or so, culminating first with the Registered Professional Engineer (PE) being a requirement for certain jobs such as bridge or building construction. In safety applications, the PE certification is required for approval of certain design considerations of scaffolds or excavations. More recently, Certified Safety Professional (CSP) or Associate in Risk Management (ARM) has been required for certain jobs in the past decade or two.

What is the occupation of a "safety professional"? Bill Tarrants, president of ASSE from 1977 to 1978, probably said it best in an article in the *American Industrial Hygiene Association Journal* dating back to July 1977 (Tarrants, 1977). Notice the blend of the safety and engineering professional even 30 years ago

> The contemporary approach of the safety professional is the unique application of engineering design and analysis, accident problem appraisal, safety program management, environmental study and analysis, safety education and promotion, and the application of various techniques intended to motivate human behavior within acceptable limits of human performance. The safety professional's consideration of people, the safe design and operation of systems involving people, and the manner by which equipment and machinery failures and errors in human performance are analyzed and various countermeasures are prescribed and applied, requires a fundamental analytic approach that is essentially different from the more academic approaches.
>
> Training for the safety profession includes not only the engineering methods of analysis and design which stem from mathematics and the physical sciences, but also from psychology, physiology, sociology, economics, managerial practice, industrial hygiene, health physics, fire protection, human factors engineering and human relations.

His concern with before-the-fact problem analysis, his contribution to product development, and his
use of objective techniques of systems performance
appraisal makes him a vital member of the management team as he works with others to stress the scientific, the economic and the management aspects of
loss prevention programs and their contributions to
the critical functions of total system performance.

So a safety professional applies math, science, and engineering
principles to analyze hazards, including conditions and behaviors, and
then offers recommendations to control costs and prevent injuries. The
engineering professional applies principles of mathematics, chemistry,
biology, and physics, with a healthy dose of humanities and behavioral
science, to practical human needs such as food husbandry or transportation, for example.

It's extremely rare nowadays to be a craft worker for 20 years and then
become the safety expert. That person is a safety practitioner, but not a
safety professional. The latter requires school—college—and, increasingly, a
graduate degree to be a safety professional. Note here that an undergraduate finance degree holder could become a safety professional, and so could a
civil engineering undergraduate, but neither, by simple virtue of science and
math courses, assumes the title safety professional without the specialized
training, again increasingly, with accreditation by a major national body, in
our case, the Accreditation Board for Engineering and Technology, or ABET
since 2005.

The same is true for engineers. My father's family was populated by
"operating engineers," who are now known as heavy equipment operators. For some thirty years or so an engineers is a name specifically rel-
egated to the engineering profession.

The safety professional and engineer of today will sit for a privately
administered written competency exam such as the CSP or ARM for safety
and the PE for the engineer. Members of each profession will have already
taken the preliminary qualification test, namely, the Associate Safety
Professional (ASP) test or the Fundamentals of Engineering (FE) exam.

Members of each profession will pursue continuing education credits
over a period of usually five years before recertifying the competency.
Here is a consensus of what I have learned about professions as opposed
to jobs or trades.

Public trust: Not surprisingly, the public trusts a professional more
than a craftsman or even an elected official, and status comes
with community standing. For example, every code of ethics
begins similar to that of the NSPE, which states, "Engineers, in

the fulfillment of their professional duties, shall hold paramount the safety, health, and welfare of the public." What higher calling is there than being a member of a profession that values public health and safety above every other aspect of its code of ethics? In fact, every code of ethics I have ever read, including the ones I use as examples (following soon), begins with the importance of public safety and public trust.

Self-regulation: Professional societies don't ask the government how to run their affairs. In fact, just the opposite happens. The federal government so highly values professional societies such as the American National Standards Institute (ANSI) or the American Society for Testing and Materials (ASTM) that it incorporates entire standards verbatim merely by calling attention to them. This is known as *incorporation by reference*: see particularly Volume 29 of the U.S. Code of Federal Regulations, Part 1926.6, for a list of 200 or so such private professional standards involved in the field of safety that have been incorporated without the interference of Congress.

Societies of professional members decide on their own rules, training, certification, and licensure of their own members. The very most stringent conduct review is accomplished by professional peers who are likely to be fellow members most closely associated with the skills and training needed to understand the issue at hand. For example, the use of a PE's seal is highly regulated, and only members of the engineering discipline may use it.

Professionals have a near monopoly on the provision of services to the public (see Harris, Pritchard, and Rabins, 1995): This is easy to see by driving your car near a mall and looking for the advertising. Attorneys, recognized professionals, provide legal services to the public on billboards and social workers do not. Medical doctors, also recognized professionals, provide health-related services to the public, and teachers, outside of their unions, do not. And professional groups tend to be zealous of their own turf in the provision of these highly specialized services.

Professionals have a high degree of autonomy at work (see Harris, Pritchard, and Rabins, 2005): Professionals have bosses, yes, but there is less micromanagement for professionals than with crafts workers or other fields. Professionals will often have their own personal libraries of textbooks, standards, and references. They usually work independently, often as a kind of in-house contractor to a law firm or medical practice or safety consulting company. Professionals rarely have set hours to work, and most often put in far in excess of the normal 40 hours. Trends toward work at home and telecommuting are growing in deference to this autonomy.

A professional often has an individualized reading list: It's good for a dinner invitation and can even help secure a business relationship

In the next few pages, I want to discuss some good ideas that don't have any research to back them up—just years of experience that tells me I am on the right track. It's about how and why a young professional needs to be a good conversationalist, and it begins often at dinner after a long day on the road somewhere.

Dinner is a chance to relax and recap the day, and maybe plan a little of tomorrow. Dinnertime is a chance to meet new people and chat without the pressures of the day or other people standing around. In fact, when I was a lobbyist, I heard an old hand tell me that "more legislation is made at dinner than on the House floor."

In a casual setting, you are able to learn about your companions, and consider their purposes but also their passions.

Other people seem to agree. In Dale Carnegie's (1936) wildly success-ful book on day-to-day business practice, *How to Win Friends and Influence People*, he underscores that knowing something about people before tomorrow's meeting will cement the relationship. Even though the fol-lowing story is written in an old style, it is important for today. Carnegie says,

> The genial William Lyon Phelps, essayist and pro-fessor of literature at Yale, learned this lesson early in life. Phelps relates:
>
> "When I was eight years old and was spending a weekend visiting my aunt Libby Linsley at her home in Stratford on the Housatonic," he wrote in his essay *On Human Nature*, "a middle aged man called one evening, and after a polite skirmish with my aunt, he devoted his attention to me.
>
> At that time, I happened to be excited about boats, and the visitor discussed the subject in a way that seemed to me to be particularly interesting. After he left, I spoke of him with enthusiasm. What a man! My aunt informed me he was a New York lawyer, and that he cared nothing about boats—that he took not the slightest interest in the subject. "But why did he talk all the time about boats?"
>
> Because he is a gentleman. He saw you were interested in boats, and he talked about the things he knew would interest and please you. He made himself agreeable.

Carnegie says it is important to make yourself "agreeable" to business contacts, and I will use the word *interesting* in place of it. They are the same thing, really, and in the world of professional safety leaders and engineers, the business or conference dinner is a major way to find other people who share your work interests and, maybe more importantly, your nonwork interests.

And by "interesting," I do not mean being pompous—quite the contrary. I do not mean spouting off details about last night's football game—no way. Furthermore, I do not mean talking about sensitive topics that may in any way offend. I mean being proactively prepared to either open a discussion or contribute meaningfully to it. This is entirely different.

Let me side track just for a minute and assume that the reader here knows about and uses social media tools such as LinkedIn. Today, LinkedIn has more than 250 million users conversing in two dozen languages. It is designed to establish and enhance a professional's networking contacts and to locate others with similar job skills. LinkedIn is one of the most commonly used social networking media for professionals in the last decade. If you aren't a member, it's time to join; you can compare interests posted in LinkedIn.

But LinkedIn won't be enough, for example, when you get an invitation to meet the Plenary Speaker at an ASSE Professional Development Conference (PDC) one evening at dinner. This invitation will most likely come face-to-face and you'll need to think on your feet about what to say and, even more important, how to be interesting, how to contribute with opinion based on facts, or even how to simply move the conversation from topic to topic.

Indeed, the business dinner is a great way to expand your professional network. The very week I wrote this, a friend of mine captured two separate contracts and the promise of a third over a single dinner that he very carefully planned to include two major players and a minor one, all of whom had been fast friends for decades, but who were going to be in the same town for a conference. The skids were already greased when Kevin proposed to them a "working dinner." Kevin is nobody's fool.

Let's talk about conferences for a minute. To start, consider this: You'll probably attend two conferences per year: one big one with a *general* membership, for example, a civil engineer will want to attend the annual Transportation Research Board in Washington, DC, and always in January of the year—it is largest single conference in the world. The same engineer will probably also want to attend a *specialty* conference, asphalt design or traffic engineering, for example.

The safety professional will probably want to go to a general conference and also a specialty conference. I mentioned ASSE's PDC, which is held around the country in June of the year. He or she could also attend the National Safety Council's annual meeting, also held around the country, usually in October. *Specialty* conferences are many: the Construction Safety Council meets regularly to disseminate new research findings and discuss regulatory changes, which are the main foci for all the these conferences.

Attending your first PDC

Let's discuss a little about conferences and then how to get yourself ready to network, share research or anecdotal findings, and broaden your own professional horizons.

Conferences are intense during the day, less so after-hours, and are divided up into three main parts:

- *Headline speakers* in the morning, often during breakfast
- *Research and discussion sessions* during the main part of the day
    - Space is usually limited and arranged similar to classrooms.
- A concurrent *trade show* in large, open areas
    - You can talk to vendors of the newest technologies.
    - The pace at the trade show is casual and means walking for three or four hours. These are very fully packed events.
    - You'll attend three or four topical sessions during the day, grab some lunch on the fly, and hit the trade show in the afternoon.
    - But after about 6 p.m., everyone breaks up into informal groups for dinner.

After a long day, it's a time to unwind with our favorite new and old colleagues. Dinner in these impromptu groups is usually out of the conference hotel but fairly close by in a small restaurant. Even though these dinner discussions have no real agenda (thank goodness—we had very tight agendas all day long), they may move across a dozen topics in an hour, and often, they get pretty intense. For example, these conversations will usually begin very generally: they might center on how energy availability impacts their company bottom line. Colleagues compare worker compensation rates and modification factors. The discussions are rarely about politics, religion, or sports, but they will discuss and compare company new-hire policies or a regulatory proposal that affects your company.

A well-known factoid for young professionals to know about is that later in the evening, participants break up into even smaller groups and may go to a bar or just walk through the downtown. At this point, participants may share job openings and even suggest candidates to each other.

I attended the Transportation Research Board for almost 20 years, and those conferences made a big impression on this young professional. My favorite hotel for that conference in the District of Columbia was always the Washington Marriott Wardman Park. It is well over a hundred years old and carries the flavor of age and patina so well. There are dozens of private places to chat over coffee during the conference or get a meal in the Wardman. Even better, there are probably 25 small restaurants across the street. If you want to get some time with a new colleague, ask a local—the doorman or the concierge—for a restaurant recommendation. You can't go wrong.

Small restaurants are ideal for good postconference dinnertime conversations, and fortunately, all are within walking distance and most all are open late. Reservations are always suggested for the better restaurants, and most people I know are comfortable not bringing their iPads and laptops to dinner with them. These restaurants are for first-class conversation, networking, and socializing. And while a good restaurant isn't meant for celebrity watching, it does happen, and it's just a bonus of a good choice in eatery selection.

Back to "becoming interesting." When you participate in these freewheeling discussions over lunch or dinner, you have a solid way to learn what is new in the fields you move in, but you can also demonstrate that you have a grip on the wider aspects of your company's position in the world. These conversations establish your credibility as a young professional. And much more often than you think, a vice president of something important joins you half way through dinner.

To compete in the conversation, you're going to have to learn to be an avid and eclectic reader, which most college grads are not. But I can tell you that most formal policy development and discussion happen in informal places, like conference dinners. As I mentioned and as I learned, most legislation is made by colleagues meeting for dinner, stopping in hallways, or even chatting in an elevator, and not on the House floor. I know that many business deals are also made in the hotel's gym, and not at some formal negotiating table.

Let me try to prove my point about becoming an interesting person through reading and conversation; I obtained a job offer one time because I was absent-mindedly carrying my airplane book to the interview. It was Stephen Hawking's *A Brief History of Time*, fresh on the bestseller list, and it just happened to be the same book being read by the CEO who was interviewing me. We talked about my job as a quality control engineer for ten minutes, he offered me the job, and then we talked about Hawking's book for another hour. That wasn't planned, but that's my point. The CEO and I became fast friends and there was an immediate and tangible trust between us.

The conference dinner, a walk on some boardwalk, the hotel gym, even a road trip with your boss—these are places where deals are made, friendships are struck, trust is established, and careers are launched or changed.

In this year alone, I recall some challenging conversations entirely not work related—that had lingering and totally positive effects: construction regulations for fall protection, the Bureau of Land Management and its wild horse roundups, how the Easter Islanders moved 10-ton statues 11 miles without wheeled implements, how the nation's new health care system will impact small construction companies.

We are now going to discuss the supreme importance of reading *outside the workplace* and its impact on professional growth. And just so everyone knows, I do not consider substituting reading outside the workplace for watching Fox News or even CNN. These latter have devolved into shouting matches and with hugely partisan mouthpieces. You probably do need to be aware of these shows, but don't spend much time watching them regularly.

And while it isn't reading, National Public Radio (NPR) is probably still the best way to get your morning news fix while you drive to work or drive back home. NPR is current, if liberal, but NPR still has the best radio news of any out there. I have listened to it for 30 years and it keeps me current in a painless way. I strongly recommend it and even rural towns are within range of an NPR station.

The value of reading heavily cannot be understated on your journey to professional growth. Test my hypothesis: Ask somebody whose opinion you value about his or her stack of books yet to be read. I am guessing they'll have between 5 and 10 books waiting to be consumed at any given time. They'll probably have novels, biographies, technical books, and probably a pile of glossy magazines, too. And even though you can start out small and build up, remember this advice. You should have a pile of books, too.

We'll now discuss the value of reading journals and daily newspapers, then books and some classics that you may have missed in high school. This is my list generated with the help of my friends across the country who also teach young professionals at the university level. I recognize that it's just a place holder until you discover your own passions and follow them. You'll quickly see I have a bias for historical, technical, and military themes, and these have worked well for me—they just work for me, that's all—mine aren't something magic. So get started on your own list.

Journals and dailies

For current social, political, and economic outlooks on trends both in this country and in Europe, there is nothing better than *The Economist*, a bimonthly news magazine by curmudgeonly British people who still are a bit angry about the United States winning the Revolution. They say things like "he is a daft governor" and "colourful," but you need to know they've been writing this journal since well before the American Civil War, specifically 1843. *The Economist* targets educated readers, and it takes clear and strong political positions in its editorials. Hillary Clinton,

Barack Obama, and George W. Bush all read *The Economist* regularly. So should you. Get it. Read it. If I only could afford one subscription, this is it, and yes, there is an online version for those addicted to iPads.

For a U.S. outlook, I would have suggested *Newsweek* or *Time* a decade ago, but not now. Sadly, these have devolved into offering mediocre news articles by lesser known writers surrounded by way too much advertising, and now, *Newsweek* may go under entirely. Opposing both of those, I suggest reading *The Week*, a different kind of news magazine with no advertising and very small but very detailed articles by renowned writers. *The Week* is a quick read on the airplane or at the gym, and besides its compact articles on the United States, and the world, it has a page each week on the movies, good books, cars, wine and food, and very strange places to live. Try it—I promise you'll like *The Week*. My students love it.

For decades, I have had a love–hate relationship with the *WSJ*. I hate it because it's expensive and it demands time and careful attention to read. So, I collect a few days' worth and try to get through them in one sitting. That's difficult because I sometimes need to give these collected dailies two or three hours at a whack. But despite the hassles, I love the *WSJ* because there is no other daily news journal with as much insight and important news about real-time U.S. and world economic and business trends—that's probably not what you'd expect if you're new to it. But it has much more—election coverage, the drug wars, impact of social media, string theory, The Hadron collider, micro-breweries of Oregon, and much more esoterica. Every day, there is one center-bottom article on page 1 that is guaranteed to make you think and probably laugh. Their op-eds are written by the biggest names in the world. The *WSJ* has a good online edition that you get with a paid subscription so you can read it at the airport or on your iPad. Every weekend, there are 3–4 page inserts on travel, food, entertainment, cars and houses we can't afford: so very cool.

Of course, you need to read your local daily newspaper, wherever you are.

Books: The mainstay of an educated dinner guest

If you read the previously mentioned three news journals plus your local daily, you'll get an invitation to sit at my table at a conference, but you haven't earned a talking role yet. We need to move on to the heavier stuff. You need to be a *real reader* to become a *real leader*, so says my good friend Col. Dave Miller, Deputy Commandant for Leader Development at Virginia Tech. Dave and a handful of my friends across the country have offered young safety professionals and engineers some really good suggestions for broadening horizons and developing insights.

Naturally, all of the people who now offer reading suggestions have standing invitations to eat at my conference table because I know the conversation will be challenging and interesting. The discussions will push my limits, and they always have. I also happen to know that these people share my belief about becoming an interesting conversationalist as part of becoming a professional.

I asked a handful of people across the country whose perspectives of higher education are relevant and current, as you'll see. I asked them what books they'd recommend to Millenials in their own classes or in their communities. Here is just a sampling intended only to challenge the reader to start a reading list of your own, and to share it with your subordinates. Remember, my list is Appalachia-centric, but your list can center on whatever resonates with your own social groups or students or employees.

Ron Kasserman is an attorney, a firearms expert, and a local historian of Appalachian history who lives in Wheeling, West Virginia. He recommends *That Dark and Bloody River* by Alan W. Eckert (1996). "Anybody who lives in northern West Virginia, Ohio or Pennsylvania needs to read this book about the families who lead people into the unknown, figured out the difficulties and survived. These early leaders gave little towns and creeks the names we know today," says Kasserman. *Publishers Weekly* says, "The lives of notable pioneer families (Zanes, Bradys, Wetzels, Crogans), the incursions of traders, explorers, colonists, adventurers and the historic exploits of George Washington, Daniel Boone, George Rogers Clark and others intersect (in this book). And if you're a weekend car explorer, the 1700s maps are good enough to use today."

Jason Musteen is an assistant professor at the U.S. Military Academy, West Point. His doctoral dissertation was about the importance of Gibraltar during the Napoleonic Wars. He suggests Doris Kearns Goodwin's (2005) *A Team of Rivals* and says, "Lincoln's leadership of his cabinet is a great example of winning over a group of strong-willed individuals, many of whom disagreed with each other and the boss, and turning them into a solid team."

Second, Musteen recommends Michael Shaara's (1974) *Killer Angels*, saying that it "may be over-used, but it does a good job of getting into the minds of leaders who have to make hard decisions. I particularly like the examination of Joshua Chamberlain as an amateur [who was] expected to make very serious decisions at Gettysburg. I also like how Lee dealt with Stuart when his cavalry failed in its mission. It's the best example I know of how to simultaneously discipline and motivate a gifted subordinate who has failed."

Next, Musteen says, "For a real-life example of the pursuit of honor, there have been four biographies of Stephen Decatur written in the last 10 years (by James DeKay, Spencer Tucker, Robert Allison, and Leonard Guttridge). Any of them would be good for ascending professionals to read. Decatur was one of the greatest American heroes of all time as a young leader of the early U.S. Navy—the hero of both Barbary Wars and one of the heroes of the War of 1812. His exploits are epic and legendary, including a lot of exciting face-to-face combat. But at the core of everything he did was honor and glory. It appears that he lived for those two things, and guys like him defined the early navy and the early republic. Dueling was often the way to settle matters of honor in the early navy. Decatur recognized the foolishness of dueling and prohibited it in his command. Nevertheless, he was killed in a duel with another senior naval officer in 1820. It might be a good study on leadership, honor, and decision making because his life is an entertaining story." And, "Beyond the potentially boring histories of World War I, there is a book of historical fiction written by C.S. Forrester (who wrote the Horatio Hornblower naval series) called *The General*. I think it's a wonderful and sympathetic examination of a guy who got it wrong. We can learn from this, too."

"Another recommendation I have for you, and not only for your students, is Grant Hammond (2001)—*The Mind of War: John Boyd and American Security*. I recommend it because Boyd was a fighter pilot first and, while studying engineering in college, he stumbled across a theory on measuring the true effectiveness of fighter aircraft that he called 'energy maneuverability theory.' It provided mathematical and scientific analysis to prove that bigger, heavier fighters did not necessarily equate to better. In fact, he showed that U.S. fighters that were assumed to be superior to the Soviets were actually inferior. He helped design the F-15 and F-16 in response, but was largely disliked by his colleagues. He spent his retirement trying to figure out why a self-described knucklehead had discovered what none of the smart guys had. In the process, he analyzed and synthesized everything he could get his hands on to figure out how people think, using math, science, military history, philosophy, etc., but ultimately with the mind of an engineering leader. Brilliant book."

Mark Hayes is a brilliant amateur historian but pays the bills being an attorney. He says, "Al Kaltman (1998) wrote *Cigars, Whiskey and Winning: Leadership Lessons from General Ulysses S. Grant*" and quotes a reviewer: "Business management is not war, to be sure, but Grant's qualities of determination, persistence, common sense, clarity of purpose and mastery of detail without sacrifice of larger vision, are equally relevant to victory in peaceful pursuits." This is not a textbook and not theory but a series of quotations, each linking the quote with a vignette where the quote came from.

I endorse this book, too, having turned to it a number of times just for inspiration for a letter I am composing or an angle for a conversation that I know is coming soon. The book is an easy read and makes a great gift. In fact, Mark gave me his copy.

Dave Miller is Deputy Commandant for Leader Development at Virginia Tech. Miller suggests *21 Irrefutable Laws of Leadership* by John C. Maxwell (1998). "Maxwell's material isn't based on research, but that doesn't make it bad: this book is easy to read and easy to apply. It's a good airplane book and has been read by hundreds and hundreds of world leaders."

Miller adds, "Professor Stephen Prosser (2010) wrote *Servant Leadership: More Philosophy, Less Theory.* Is it a philosophy, or a theory, or a set of values, or a list of characteristics, or a series of practices—or some combination of all of these things? Professor Prosser addresses these questions in the context of the literature and research on servant leadership, which is exactly what we preach and practice here at Virginia Tech. After reviewing the ways in which people try to describe and explain leadership, he provides six reasons why servant leadership is a philosophy, not a theory, concerning service and the practice of leadership. The essay is concise, and designed for the practitioner."

"For emerging leaders, I use Greg Ballard's book, *The Ballard Rules: Small Unit Leadership,*" says Miller. "Greg is a retired Marine and currently the mayor of Indianapolis. His book is short, easy to read, concise and full of wisdom. It works for any junior-level professional.

Bob Hayes was the president of Marshall University and still provides educational consulting today in his late 80s. He recommends *The Case for Servant Leaderhip* by Dr. Kent M. Keith (1998), Greenleaf Center CEO. "In this 85-page book, the author argues that servant leadership is ethical, practical, and meaningful. He cites the universal importance of service, defines servant leadership, compares the power model of leadership with the service model, describes some key practices of servant-leaders, and explores the meaningful lives of servant-leaders. The book includes questions for reflection and discussion."

Casey Brower whom we met earlier as he talked about the importance of self-awareness, is a historian and academic leader at VMI. In conversation with me, Brower recommends *'In Extremis Leadership': Leading as if Your Life Depended On It* by Col. Thomas Kolditz (2007), U.S. Military Academy. "When there are no 'do-overs,' you need leaders who can learn rapidly in situations that are not in the textbook. These leaders live a common lifestyle with followers and share the risk with them. These factors encourage followers to trust and show higher levels of loyalty than with other models of leadership. This book is maybe the best single book on how to spot leaders who can do exceedingly difficult tasks under the most challenging, emerging conditions. This is a conversation starter anywhere among safety people or engineers."

I have put my own recommendations at the end of this chapter in an attempt to tie all of these together. Remember, even though this *book* is about leadership, this *section* is about encouraging young professionals to become interesting professionals in their own right.

I am batting 1000 with the following book: Every student who has read it has thanked me for the referral. *At Home in the Heart of Appalachia* was the first book written by John O'Brien (2001), a native West Virginian who was observant of the human condition at many levels. When his rural community came face to face with an institute brought in from out of state to take over the fire department and then the school system, the normally independent locals fought back when it was obvious that these do-gooders simply thought they were smarter than the locals. This is "toxic leadership" at its most deadly, and along the way, O'Brien spins a narrative of condescension and stereotypes that the institute foisted privately on local citizens. I treasure my copy of this book and I have squirreled away a supply on my shelf to give my friends.

Moving along, here is a young man of 23, known worldwide today, but unremarkable in 1755. He was working hard to become a leader but failed in his first effort, which ended with almost 50 of his subordinates killed. I recommend *Drums in the Forest: Decision at the Forks and Defense in the Wilderness* by James et al. (2005) because most people today don't know about the early years of George Washington's life and how he deliberately overcame adversity by addressing his own leadership failures point by point. The book also illustrates how arrogance and toxic leadership of even the most famous heroes can cause their demise when they have to play by rules established outside their comfort zone (Figure 3.1).

I have a room in my house dedicated to the memory of a man who lived long ago in my general area but was an enigma to his friends. He was a struggling student, had few friends, and talked to himself in difficult moments, dressed not in colors and flamboyance but the same as people who worked for him. And like Washington, he was out to learn

Figure 3.1 The battle of Fort Necessity was George Washington's first test, and he lost. Find out why by reading about it.

from his mistakes. I have a dozen or more books about T.J. Jackson, but none so poignant as the obscure one I found in a used bookstore in rural Arkansas one summer.

In *Lost Victories: The Military Genius of Stonewall Jackson* by Bevin Alexander, I found in case after case original research suggesting that Jackson was the genius behind the Confederate strategy, not Lee as most people have always assumed. Most importantly, after each and every defeat or victory, Jackson dissected which leadership strategies worked and which did not and changed his tactics accordingly. When the call came, Jackson led his own students to defend Richmond, and for two more years, he "mystified, mislead and surprised" far larger armies and much more famous generals. In my view, he was the most humble man in the most trying of times. How he painstakingly learned from him has changed my own life (Figure 3.2).

Early in WWII, George Marshall wrote a personal letter to *every* mother who lost a son in the war. After the war, he made sure that each of his staff generals had a parade of their own when they returned, even while denying himself any such notoriety or fluff. During the same time, Marshall brought the U.S. Army from 19th in size at Pearl Harbor (we were smaller than Portugal's and Bulgaria's armies) to the world's best after about a year and a half (*Soldier Statesman Peacemaker: Leadership Lessons From George C. Marshall,* written by the American Management Association, but there are a lot of similarly good books about Marshall).

Figure 3.2 Dr. Winn's Stonewall Jackson Memorial Library.

Figure 3.3 Most American generals demanded their own ticker tape parade when they arrived in the United States after WWII. As the ranking general, George Marshall ordered parades for everyone … everybody but himself.

More than anything else, Marshall is remembered for a speech that he said would be "a few words … maybe a little more" at Harvard in 1947, where he already humbly turned down two honorary doctorates. Instead of seeking accolades, he gave a low-key presentation that launched the Marshall Plan, saving Europe from starvation and probably the rest of the world from World War III. Like Jackson, Marshall loathed ego but valued loyalty, and he took care of his people first. After Jackson, Marshall is my main American leader and hero (Figure 3.3).

There's nothing to say you shouldn't have your own list of favorite inspiring books. Indeed, you should have your own list and add to it every chance you get. And you can pay it forward—there's nothing better than giving a young person a book that you have inscribed, a book that moved you along your own journey to being a better, stronger person.

Biographies, histories, and technical books all keep your mind sharp and serve to inspire and provide substance for engaging conversation. They allow comparisons to the past. They let us see what historical figures did in situations young people will surely face in the future. They teach us about humility, self-awareness, and personal courage. They teach us that, sometimes, we have to dig deep inside ourselves and to bite our tongue. And they show us how to teach these things to our own subordinates.

chapter four

Further becoming a professional

Dr. Winn's 50-plus time-tested rules for professional success: Managing your time and office

I wish I could say that there is empirical research here from which I could draw examples about becoming a professional safety leader or engineer. Because I couldn't find any, I drew on my own experience and developed some rules about professional success. I hope the reader finds these anecdotes useful.

Here are some things to think about. First, being a graduate student is a full-time job. Starting out as a young professional requires even *more* time. Entry-level safety professionals and project engineers spend 40 hours minimum at work; more typical is 50–55 hours. At the far end of the spectrum, Kevin H. was putting in sometimes 90 hours a week in a local mining industry and he did it for almost three years. His main conclusion was this: "You have to be organized. You can do it, but organization is the key."

Here are some suggestions about managing time and your own office.

- It will take about a year to master the basic requirements of your job: Be patient. These simple but important administrative tasks include how to process paperwork for travel and purchases, how to communicate from one office to the other, use of credit cards, where to get an approved rental car, and so forth. Just be patient.
- Keep records of projects you undertake and try to keep a running portfolio with photos and notes and drawings of major accomplishments. These will help you in performance reviews and you can show your work if you apply for a new job internally or externally. I wish I had copies of the best work I have produced over the years.
- Don't be intimidated by a bureaucratic and slow moving system; normally, it is designed to help and protect you if you work within it, not against it. The larger the group, the slower the bureaucracy.
- As you begin to understand your position as a professional, recognize that it may be your first experience with real layers of bureaucracy and a line organization. Be patient understanding it and who reports to whom.

- Make sure you know the formal and informal power structures: Ask peers. Ironically, the informal power structure is the stronger of the two.
- Get familiar with your company official system. Get the formal rules down. There are usually policy manuals available. Sooner or later, you're going to need it.
- Knowing names counts for a lot. Knowing first names counts for even more. If you have subordinates, learn everything you can about them, but as a general rule, don't "friend" them on Facebook.

Get your own organizational tools:

- Get a good pad-folio with an embossed logo such as a university logo or company logo, and not a sports logo. Carry it in Dr. Winn's prescribed way: under your left arm, open at the top. Every time you go into your boss's office, carry it with you.
- Carry your pad or supplies with your left hand but shake hands with your right hand whether you are right-handed or not. Most people are right handed, and this eliminates the awkwardness and fumbling of wondering.
- In addition to the pad-folio, carry the necessary pencils, iPhone or iPad, and previous notes to your boss's office and do this for *every* meeting.
- Get a day and week planner; they are $6.00–$15.00 at the bookstore or Barnes and Noble. Your iPhone has a good calendar, too, and I use mine, but the hard copy becomes more of a companion like a good book does.
- Schedule your major activities in the planner (i.e., meetings, vacations) or your iPhone or iPad.
- Select and use only one, single name brand pen and one, single name brand mechanical pencil. They become your good friends because they are predictable. Value them. Toss the junk pens and pencils now or give them away. Don't loan your favorite pens and pencils. I recommend Cross pens because they are reasonable to purchase, and because they are dead reliable. As a bonus, you can have them rebuild periodically because they have a lifetime warranty.
- Get a good calculator. I recommend a Texas Instruments because of the predictable notation and input. Use the same one you graduated with because you are familiar with it in every way. It is your friend. Cherish it. Don't loan your favorite calculator.
- Your iPhone or iPad can also work as a calculator, but be aware of notation nuances.
- Electronic data storage has moved from zip drives to flash drives and surely will evolve further. Just try to keep these organized; you

can use nice leather multipocket carriers that are really convenient to travel with. I am not good at this, but I am working on it; I think I have about 35 flash drives in various places.

- Each project worthy of presentation needs a folder or nice cover for it. I have a supply of university-logo color paper folders just to carry things around in and I give them to students for the same reason. A nice color folder with an embossed logo costs only $1.99 and it makes subordinates feel good when you hand them a folder for use there in the office or on the road.

- It has been my experience that a seasoned professional would rather have a small gift of almost anything with his alma mater's logo on it, as opposed to a mug or something with an industry logo. I keep a supply of small, high-quality lapel logos in my office and hand them out liberally when the alums are in town. They are like gold and inexpensive, too.

- You ought to order a couple of packages of yellow tablets (not legal size) for your own office. I go through a box of them per year (about 40) and you probably will, too.

- Back everything up on your computer about once a week or so, or get an automatic back-up system through your organization. It's a hard lesson to learn later when hard-fought spreadsheets evaporate in a puff of smoke some day.

- Cloud-based data storage is here to stay. Use a G-drive or iCloud for information you need access to regularly, especially on the road.

- Watch what your new-hires are using; they'll have the latest apps, the latest phones, the newest softwares, the latest e-storage ideas. When I have software issues, I can call our IT folks or find a third-year engineering student. Guess which one I pick.

Some simple things pay real dividends:

- Be early for every meeting: The correct arrival time is five to six minutes prior and not 10 minutes prior. If you're more than 10 minutes late for a nonimperative meeting, consider not going at all and sparing yourself the embarrassment of being a nonachiever.

- If you do end up as a faculty member somewhere, amaze all of your colleagues and try something really new: be on time for meetings.

- Be prepared for the day's meeting discussion; read ahead; check your notes from the last meeting. You can train your own subordinates in the Thayer method used back in the 1800s, which meant that every student was prepared to actually teach the next day's material. That meant every student had to be prepared for the day's discussions.

- Know how to knock on a door and enter a room in "Dr. Winn's pre-scribed way." Stand outside. Two or three knocks with your middle knuckle on the door if closed or door frame if the door is open. Do not enter until your presence is recognized. Smile, enter, and don't sit down until you are offered a seat. Takes notes and file them for next meeting. Do not overstay your welcome.
- Arrival time at work is always 10 to 15 minutes before that of your boss. Departure time is less important, but at least be there before the boss arrives.
- If you have a field-type job in construction or oil and gas, be prepared with your own "five-gallon bucket" in your truck bed or behind the driver's seat. This means being ready for some of the day's needs: maybe a tape measure, a sting line, even abrasive disks, or saw blades—you call it. All of these eliminate a drive back to the field office when your new hires forget their own supplies—this wastes time and tries your patience.

Erik Edwards is a young civil engineer responsible for erecting steel, pouring concrete, and applying metal roofing. He is also responsible for the safety of his work crews. "Once I started my position with a big construction company, I realized Dr. Winn was right about being ready for surprises every day, especially in construction. I followed his advice and got my own five gallon bucket. Here is a photo of it with my hard hat and a tie. I have a couple of hammers, screwdrivers, chisels, wire brush, and grinding disks for angle grinders—the kind of thing that craftspeople are hard on anyway. I just hand out the things in the bucket to save time and prevent using dull tools or used-up disks. They can shatter violently. I also have a digital voltmeter, a laser level, and a copy of the Code of Federal Regulations 1926, the rules for construction safety. Recently, I added a clean shirt and tie for unscheduled owners meetings (I did put those in a toolbox and not the bucket)," Erik said. "My bucket is a simple way to prevent a lot of surprises" (Figure 4.1).

Tips for successful meetings:

- Never, ever let your phone or tablet go off in a meeting. Otherwise, you will soon be permanently branded as "that guy."
- If something is unclear, be sure to ask the boss at the end of that meeting. Try not to interrupt during the meeting unless it's imperative.
- Be aware that *operations meetings* are usually every day. Be prepared. I have mentioned this anecdote, but it's important, so here it is again. At one manufacturing job I had where I was the industrial engineer, production meetings were brutal—nobody was allowed to sit down, and every management team person had his or her five minutes

Figure 4.1 I have often suggested, and only half-jokingly, that it's a good idea to put a five-gallon bucket behind your truck seat to carry the irregularly used essentials: an extra hard hat, some hand cleaner, a wire brush, some grinding disks, and so forth, and of course, a tie for a last-minute meeting with boss. This is Erick's bucket.

under the white heat of fear that the boss wasn't going to be happy. Be prepared to respond.

- Take notes at every meeting, but don't try to write in sentence form when an abbreviation is probably OK. Use the same abbreviation system you used in college, and then file your notes somewhere where you can find them again easily.
- Listen for cues in meetings: High pitched or loud words and repeated words are cues that this concept is important.
- If the boss is looking at you or if the speaker uses numerical examples, it's probably important. Take notes.
- The last few words of the sentence are usually important. Train your brain to listen for that kind of nuance.
- Keep your notes in a loose-leaf binder with no ragged edges. This way, you can move pages or add handouts all you want. I have a

good friend, Ed. Y., who has filed this kind of notes for over 20 years, and yes, he can find things that far back, too.

- Get a small, flat three-hole punch for meeting handouts so that you can put them in your binder.
- Summarize and rewrite notes around the end of every week. This is hard to find time for, but it pays dividends because most people just pitch the notes.
- Not everyone will agree with this advice, but I recommend that you don't share notes with people who make no effort to take them on their own. These are the ever-famous "nonachievers." Sorry, sometimes life is cutthroat.
- Sit in the first couple of rows of the meeting room or directly across from the boss. You'll hear everything and the boss will know that you are serious about meeting. The more important the meeting, the closer to the front you should sit. Back-row sitters are still, and always have been, nonachievers.
- If you know you will miss a meeting for a doctor visit, e-mail or (if you are comfortable with it) text the boss ahead of time. Don't surprise the boss by being absent at an operations meeting.

Your employees and "management by walking around" (MBWA):

- Soon enough, people will work for you. As Collins (2001) says in *Good to Great*, "get the right people on the bus and get the wrong people off the bus as soon as you can." I discuss this further in the book, but suffice it to say that unproductive whiners must be ushered out as soon as you can do so.
- Make a point to speak to every employee that you possibly can at least once per day.
- Start MBWA in Dr. Winn's prescribed manner; it starts about 7:30 a.m. You already know employee names, so you can exchange informalities just walking by. You also learn where manpower holes will be that day (absenteeism) and where problems still lurk from third shift, and which problems are now handed off to you.
- Smiles are indispensable tools of MBWA, and they are still fairly inexpensive.
- MBWA is usually pretty brief. In the morning, if you are at a manufacturing plant, arrive before the shift, get a cup of coffee, and do your MBWA until the coffee is finished; this is the caffeine version of MBWA, but it works.
- Often, your early-day MBWA locates operational problems before the department heads do. As a young leader, you can evaluate the problem and propose a solution even before the others arrive. This preemptive action makes you simply invaluable.

- If you have an Employee Stock Ownership Plan (ESOP), post your stock quotes in your office and update them every day. I learned this trick from a friend, and he tells me that it has stimulated conversation and understanding of the company's loss control function for 20 years.
- Know what your own incidence rate and days away from work numbers are and post them, too, and tell employees why the numbers are important. Put them on the same dry-erase board as your stock quotes; same reason and benefits.
- Drag your boss and, later, your boss's boss down to your work area and do regular walk-arounds while you meet employees and talk about safety issues. Naturally, all of your own bosses need to know you as the chief engineer or safety professional on sight. Upper management visibility is critical to your own credibility in either capacity.
- Be nice to the people who aren't in your line of direct reports. They'll appreciate it, and they'll be more likely to help you or be more helpful when you really need it. This includes the department secretary, the lab assistant, the purchasing agent, the parking attendant, the security person, all of them. Most of these folks know the best places to fish, too—you can share information on nonwork items.
- Know the maintenance crafts people and any shop foreman. Often on your early morning MBWA walk-around, you can spot broken equipment or something that fell off overnight on the back-shift and get it fixed without a work order by just calling the maintenance crew.
- Do not just assume in the foregoing that the maintenance crew will help you in off-hours. They may help you fix your lawn mower, but show your appreciation with a gift card to Lowes once in a while or a six pack that you leave in their truck bed.
- Once again, about the informal power structure, even though the maintenance crew pretty much all have welders and shop tools, don't overdo it asking them to help you. These guys are craftsmen, so respect their time before you ask them to fix your tree stand or cabinet hinge every week. They'll begin to resent it.
- Never consider having alcohol at lunch or on the road with subordinates even if others may offer. The positives are few and fleeting, and the negatives are huge and they remain.

Resources for yourself and your office:

- You will probably need specialized drawing or accounting software (Autocad, ProEngineer, Solidworks, Athena, or others specific to your industry). Your employer should be willing to get them for you—don't be afraid to ask for them.

- You will need certain textbooks or references (National Fire Protection Association [NFPA] codes, International Electrical Code [IEC] codes, International Code Council [ICC] codes; American Society for Testing and Materials [ASTM], The American National Standards Institute [ANSI], Occupational Safety and Health Administration [OSHA] codes, the National Safety Council series of books, and others specific to your industry). Once again, don't be afraid to ask for them.
- Ask for other materials you need; you will have a budget for these purchases.

Travel: You're going to do a lot of it.

- You'll have a travel budget and credit card. Never, ever abuse them with private purchases or weird things. Plan to use your own money for the weird things so you don't have to justify it later.
- Don't abuse your per diem (daily travel allowance). A good friend of mine lost his position as a high-level, national board president because he thought nobody would care if he chiseled the company on travel per diems. He ruined his own reputation for about $20 a month.
- Know company rules on alcohol purchase (most state employers prohibit alcohol purchases on any receipt). My simple advice: Don't.
- Know your company rules on hospitality (buying lunch for workers or vendors) because sometimes this creates uncomfortable and potential conflicts of interest. Better to say "three checks, please" when you order your meal and head off any future issues.
- Watch others when you order: Don't order the filet mignon every time.
- Never, never drink too much on company travel among peers or your boss. You will be remembered for years as "the guy who drank too much," and the reputation will follow you.
- If your host wants to refill your glass too much, it's OK to cover your glass with palm, or else pour most of it down the sink later.
- If you do drink, go light or ask the bartender for water and a lime. It looks like a mixed drink. I have done this many times.
- Be aware that a lot of travel often means gaining weight. You can ask for hotels with gyms or work rooms. On the other hand, I have friends who do a lot of personal body building by choosing the best hotels with good gyms.
- Experienced travelers often pack their own pillow and good ear plugs.
- On the plane, a good neck pillow is a must for a lot of people.

- People do weird and wonderful things on the airplane; be patient when the person in front of you puts his or her seat back into your laptop.
- If you get stranded or bumped for your flight, you can change your flights while standing in line if you have a smart phone or iPad—pretty cool. It saves a ton of time standing in the ticket line behind irate dads going to the beach with their kids.
- Ask your company's human resources people about rental car insurance. It's complex and can be personally disastrous for you as an individual if something bad happens. Protect yourself by knowing the rules. Same goes for lost credit cards, both personal and company cards: Might you be liable?
- Lots of companies allow professionals to telecommute. This means it's OK to work at home a couple of days a month. Don't abuse the privilege by getting caught on the golf course on your telecommuting day. But then, I have always said that golf is its own punishment anyway, so you get punished twice this way.

Make the quick transition from student to professional: Five easy rules

So we have now discussed what constitutes a profession, and how they self-regulate and require certification tests periodically. We discussed that professionals usually don't have set hours but are on the job in excess of the normal 40-hour week. A student in safety or engineering should already be studying to become certified as a CSP or PE, most likely, but these things take time. What else can a person do to quicken the pace of becoming a pro? These are my five easy rules so you can think of yourself as a pro starting today.

1. *Outside reading* is a way to study trends in your field and develop global interests. You will be expected to be aware of current events and economic trends. You've heard me say it already earlier in the book, but to reiterate, this is hugely important. I recommend at least one daily newspaper, one weekly news magazine, and regular professional forums via the Internet. The pros read voraciously in a half dozen fields. Start with my book list in Chapter 3 and add your own.
2. *Dress the part:* Never wear jeans with holes at work or even on travel; shirts always tucked in. No baseball caps indoors; polo shirts are replacing shirt and ties, so that's OK. For girls, showing less skin is better under every circumstance.
3. *Act the part:* Carry a good-quality ballpoint pen and pad-folio and take notes; leave the spit cup outside along with the sports logo

ball cap. Know how to knock on doors and enter a room; study etiquette and protocol for your office/or industry and if you aren't sure, ask (for example, are muddy boots OK to wear in the construction trailer?). Avoid office politics for a year after you hire on. Men, a good tie won't kill you. Women, the less skin the better. I am reiterating some of no. 2 because it is important.

4. *Join professional organizations:* The American Society for Safety Engineers, the NSPE, The Society for Women Engineers, the ASEE, and The American Society for Engineering Education are all starters. Most professionals in safety or engineering are members of six or seven such societies after working four of five years in the field. This is a way to learn about trends and special technical seminars. And, of course, these conferences are the single best way to network.

5. *Immediately begin to seek professional certifications:* The Graduate Safety Professional (GSP) or ASP will precede the CSP by four years or five years, respectively, and for engineers, the FE will precede the Registered PE certification by five years. Other possibilities include the ARM and the Certified Industrial Hygienist (CIH). Any certification worth having will require a preliminary qualification test, periodic recertification, and continuing education credits. Don't just "buy" a certification online. Work toward one that's recognized in your field and start working for it the day after graduation.

Do I live this with my own students? You bet I do. In lecture no. 2 after I introduce my Management Principles graduate course, I begin to stress the importance of the transition from student to professional, and from professional to leader. I took points no. 1–5 from lecture 2.

Leading after managing: It's the future

Now that we have thoroughly discussed the path from student to young professional, let's talk about what happens after that. As I have laid out in Chapter 1, the generalized mission of safety professionals (compliance and program maintenance) isn't enough to aspire to anymore. Consider the definition of *management* itself (planning the activity; organizing materials, timing, and resources; leading people in the accomplishment of the activity; controlling through the application of reward and punishment and its variants; and evaluating the success of the activity). This allows no room for higher goals and assimilation of a company's core values. Striving for higher goals to create a vision for the future, and not concentrating on ways to make more widgets—this requires leaders above managers.

I will make the case as we go through the following material that leader development is easiest in a supportive company where leaders are identified early and trained continually. I will also make the case that leader development is also possible in a microenvironment (a department, for example) or an otherwise unsupportive environment where upper management doesn't seem to care about leader development. What an authentic leader does can still be practiced in these "depleted" kinds of environments.

Academic safety units are recognizing that leadership is going to be more important than management in the future, as I have shown in Chapter 1, where industry is demanding leaders. In recognition of the fact that management only goes so far toward working in values-driven safety and engineering cultures, some academic safety programs (the WVU program, for example) have modified their central missions in 2010 to be *"developing leaders* to preserve and protect the people, property and efficacy of an organization."

The change is subtle but very important. The change integrates people with the core values of the company; it asks a leader to step up on behalf of his people—to take extraordinary steps—when the job description of a "manager" would say, "stay put." Leaders are recognized as agents of change, whereas managers may or may not be.

There are only a few variations in the definition of *manager*, but there seem to be endless definitions of what a leader does. We can say that despite the myriad definitions of *leader*, we conclude that he or she must be a good manager first, by default, and set about taking care of people and the organization's unmet needs. Remember, a leader does the right things, while a manger does things right.

And because a leader in safety management or in engineering can be called upon to act under life-threatening conditions—as Collins (2001) says, "people can die" when these leaders make a mistake—safety leaders and engineering leaders must be prepared for more and tougher challenges than managers of the past. Leaders will be hired to work in far-reaching jobs across the world; work in a variety of cultures, social norms, and more; to speak different languages; and to be more interconnected than ever before.

Leaders will be called upon to clarify missions, to set strategies, and to inspire and energize people in ways that managers are not prepared to do. As today's leaders move among business, nonprofit, and government and across countries, authentic leaders are learning to speak a common language. This is the language of organizational vision, values and mission, and strategies based on them and of serving the needs of the customer, no matter what the organization's goal, product, or service.

What others say about the importance of leadership

Here are some examples of how current theorists in leader development define what a leader does and why leaders will be required in the future of safety management and engineering. Some are quite simplified, but maybe that's the point (Figure 4.2).

- Frances Hesselbein and Rob Johnston (2002), editors, *On Mission and Leadership*. "Exemplary leadership involves five practices: model the way; inspire a shared vision; challenge the process; enable others to act; encourage the heart."
- James Kouses and Barry Posner (2007), *The Leadership Challenge*. "Jack Welch [long time CEO of General Electric] says he had only three jobs at GE: selecting the right people, allocating capital resources, and spreading ideas quickly. Without leaders who can attract and retain talent, manage knowledge, and unblock people's capacity to adapt and innovate, an organization's future is in jeopardy."
- Warren Bennis (2002), noted author on leadership, *On Mission and Leadership*. "Timing is almost everything. Knowing when to introduce an initiative, when to go before one's constituents—and when to hold off—is a crucial skill. Leadership is about building connections. Effective leaders make people feel they have a stake in common problems."
- Doris Kearns Goodwin (1998), writer and historian, *Leader to Leader*. "People want direction. They want to be given challenging tasks, training in how to accomplish them, and the resources necessary to do them well. Then they want to be left alone to do the job. A leader does that."

If Patton himself could not define leadership, can we?

Even world-recognized leaders sometimes have trouble making concrete what they know about leadership itself. General George Patton said this in a letter to his son toward the end of WWII, well after his credentials were established as one of the best generals in history:

> Leadership... is the thing that wins battles. I have it—but I'll be damned if I can define it. Probably it consists of knowing what you want to do and then doing it and getting mad if anyone steps in the way. Self-confidence and leadership are twin brothers. (Connelly, 2002)

Patton was modest, of course, but he was best known for the very traits we will soon discover about leadership models: sharing risk with subordinates, self-confidence, and leading from the front.

Figure 4.2 Leadership has many definitions.

- Frances Hesselbein and Eric Shinsecki (2004), coauthors of *The Army Leadership Field Manual.* "A leader who is self-effacing and lacks charisma may fail to inspire confidence. A charismatic leader who believes that he or she is more important than others will eventually lose followers. To inspire both loyalty and excitement, a leader needs to couple humility and charisma. Both can be developed through reflection, feedback, and an emphasis on authenticity."
- Patrick Leoncioni (2002), widely read author on team management, *On Mission and Leadership.* "Leaders must communicate their organization's mission to all parts of the organization. The mission provides a reference point, an anchor, and a source of hope in times of change. When it connects with people's values, it brings purpose and meaning to those who are fulfilling the mission and provides the impetus for creativity, productivity and quality in the work and in personal development."

We can easily see from reading these highly varied definitions of leadership that there is little real commonality, but some themes emerge. There is repeated use of certain constructs and words such as *values, communication, mission, ethics, personal development,* and *management of talent.* As simply as I can put it after sifting through dozens of authors who are represented in small part by the citations in the list, a leader does more: a manager takes care of the company and its business, whereas a *leader takes care of its people.* Now let's learn what that means, recognizing that the elusive notion of leadership has been lost on even some well-known leaders, including General George Patton of WWII fame.

As I have mentioned, the main mistake I made when I began collecting material to instruct an academic course in practical leadership was that we could just start with leadership, define it carefully, and go backward to its underlying causes and influences. In fact, the opposite is true, but it took some late-night reading and intense personal interviews up and down the East Coast to make sure I got the point, which is this: A potential leader has to start with *self-awareness* (review Figure 2.1 again to understand this progression). *Personal self-assessments* cause us to be introspective ("what do I believe about myself?"), and they are a way to elucidate drivers and motivators inside a person. The best of these include the academic type, such as the Myers-Briggs or Real Colors, which are discussed further in this chapter.

After an assessment of motivators, and only after it, comes an evaluation of personal *core values,* or an exploration of self to the world. Core values ("what do I believe about my relations with others in my life?") are the very roadmap to the creation of leadership: You just can't jump straight, to leadership without a candid self-assessment of what you believe about trust in other people, or even your own belief in a deity. It's an examination of your affiliations and your ability to influence them.

It may be intuitive that self-awareness leads to establishing core values, which in turn leads to establishing organizational values. An authentic leader or an authentic organization expresses and *acts* upon a systematic exploration of value systems that end up consistent with their behaviors. Let's examine this system of awareness supporting values, in turn supporting behaviors, which in turn support authenticity of leadership.

Let's define *leadership* more formally now: a process whereby an individual influences a group of individuals to achieve a common goal. This is a definition widely used and attributed to Peter Northhouse, and even though other definitions are out there, this definition makes some implications that are useful to us. Foremost, it suggests that leadership is not something you're born with, although sometimes it is difficult to not say, "that person is a born leader." The vast and overwhelming majority of research and literature available says that leaders can be made through the same process we discuss here in this book: They size themselves up first, declare their own core values, work hard to transfer them to the organization, and then train others and themselves to abide by them.

But even latent leaders may be faced with events that challenge others to fall back. This happens particularly when the dire circumstances showcase the person's strengths, say when he or she adapts successfully in times of emergency. Northhouse and many, many others, including military and industry leaders, suggest that *leadership can be emulated and learned, but it isn't genetic. Leadership is therefore a learning process.*

As Northhouse says, leadership isn't a trait by itself, but a set of traits that we can refine but not something we are born with. If leadership were something only certain people were both with, there would only be a certain number of leaders possible.

This "fixed amount" model would cause people to throw up their hands at the idea of becoming a leader.

How, then, did Jack Welch and Colin Powell get to the top of their fields? They will tell you themselves that they used the innate traits they did have to *choose to become a leader.* They were born with attributes that Smith and Ford (1998, in Northhouse) say are important to the development of leadership; these attributes are dominance, intelligence, and confidence.

These attributes, or traits, are necessary but not sufficient. They merely set the stage for leadership to be learned, and without them, learning leadership skills is much less likely to be successful.

What about charisma, does it underlie leadership? My first response would be "no" because charisma is a trait, something we are born with.

chapter five

Core values underlie leadership

I am making the case that leader development is going to be a requirement for many of the best of the next generation, not just a luxury, and that leader development requires a choice and some effort. The investment in time is spent on developing into a professional and later on personal awareness, which we have discussed in Chapters 1–4. Only after that and a choice is made to become a leader, time is invested analyzing personal and, later, organizational values.

I am saddened when I see a company's quarterly report and early in the report is a glossy page with "Our Core Values," which have obviously been constructed by a committee intent on pleasing everyone. Every hot button is touched, every minority is addressed, every union accounted for, retirees are happy. "Everybody is happy," and that's our set of core values.

This isn't the right way for an organization to establish its core values— they must come from committed and motivated individuals who shape their own values into something the organization can agree to, not just be assigned. Personal values of the organization's leaders become the very core set of values that the organization follows.

An organization's core values aren't assigned to employees by a committee: They come from individuals who have examined their own motivations

Core values are principles to fall back upon—a source of strength—to guide the future when things are going well and a safe harbor when things go badly. That is the topic of discussion now: what core values represent first to the individual and then to the organization, and how to learn what they are.

Core values

A value is a construct, a principle, or standard that is worthy for its own sake and needs no explanation. These are beliefs that guide personal behavior whether at work and elsewhere and do not require external validation. They specify what we stand for and clarify what we think is important.

Core values are not business strategies or decision algorithms, but they may underlie decisions. They are *prescriptive* (what we should do) and not *normative* (what we actually do). Core values represent our higher ideals.

Examples are democratic core values as described in the Declaration of Independence (shown in the following), and please note that these are not dependent on some system of court decision or other validation; they exist because they are universally understood:

> We hold these truths to be self-evident, that all men are created equal, that they are endowed by their Creator with certain unalienable Rights, that among these are Life, Liberty and the pursuit of Happiness.— That to secure these rights, Governments are instituted among Men, deriving their just powers from the consent of the governed,—That whenever any Form of Government becomes destructive of these ends, it is the Right of the People to alter or to abolish it, and to institute new Government, laying its foundation on such principles and organizing its powers in such form, as to them shall seem most likely to effect their Safety and Happiness. (Excerpted from Paragraph 2 of the Declaration of Independence)

Core values can be held by an individual as in a personal manifesto (a statement of what you hold truly dear) or a company can express them for itself. Note that in the Declaration of Independence excerpt, the writer shows what the new country stands for, what is important, and that no external source of validation is necessary. The document didn't try to be all things to all people, but it did seek the higher ideals of people everywhere and not just those in the 13 colonies. In fact, over two dozen other countries across the world have borrowed this text and even some of the statements because they are universal.

Research suggests that core values are learned early in life from parents and extended family, church leaders, key teachers, and even sports coaches. Individual core values, once in place, are also difficult to change.

Why do we need core values?

The U.S. Marine Corps distributes little cards with its three core values printed on them as a reminder to make them personal and public—to say to the world, "this is what we stand for." The other service branches have similar reminders in the form of posters and leaflets. Increasingly, larger industries are adopting core values to declare publically their central views of themselves.

Parsons Corporation, one of the largest constructors of airports, buildings, and water systems in the world, employs close to a thousand safety and engineering professionals, including the former Vice President of Safety and former WVU football star, Andy Peters. While at Parsons, Peters helped transform the company's safety operation from a compliance-based function to one based on respect for the individual and personal responsibility. After a time, core values at Parsons' safety operation rose to the highest levels of what the company stood for— Parsons itself names its core values as safety, integrity, innovation, quality, diversity and sustainability, and Peters says these values still drive the development of the *entire company* at all levels.

Even a university can declare core values. Col. Dave Miller, deputy commandant for leader development at the Virginia Tech Corps of Cadets and director of the Rice Center for Leader Development, writes about Tech's massive limestone sculptures called "pylons," which represent the entire school's core values. Col. Miller writes,

> The pylons in Figure 5.1 embody the values that we, as VT students, faculty and alums hold dearest. Each sculpture is part of a story that begins with the woman and son, representing the alma mater and student. She tells her son that if he lives up to each of these values, he will lead a good and exemplary life. Every VT student is taught these values and the story.

Virginia Tech Magazine continues the story of the much revered pylons at Tech (Figure 5.1):

> Although War Memorial Chapel, completed in 1960, was initially intended to honor Techmen killed in World War II, the names of alumni who have died in military conflicts beginning with World War I are now carved on the Pylons. The majestic Pylons stood watch over you as a Virginia Tech student, and they still do. The names of the Pylons—Brotherhood, Honor, Leadership, Loyalty, Service, Sacrifice, Duty, and *Ut Prosim* (That I May Serve)—embody the values that members of the Hokie Nation hold in highest regard.

If an individual can identify core values, the individual can then look to see if his or her actions are congruent, and if not, work to make them so. By identifying his or her own core values, a leader can understand what drives other people and finds ways to connect with those possibly

Figure 5.1 The huge pylons at Virginia Tech embody the values that are shared by all faculty, students, and alumni.

opposing core values. There are complex and there are simplified ways to identify a person's core values and those of subordinates. Of course, in history, recognized leaders did not have the advantage of a written assessment instrument but nevertheless must have gone through introspection instead. Among many others, Frederick the Great was recognized as a voracious reader and author as a teenager. His writings reflect deep introspection about his own motivations. But for us, an understanding of self-awareness can come in many forms, including paper-and-pencil instruments.

One of the most widely used values inventories is the Myers-Briggs Personality Inventory (MBPI), whose use dates back six decades to WWII, when it was used to identify personality types most suitable for a particular kind of war-related occupation. The Myers-Briggs website (Myers-Briggs.org) says:

> The purpose of the Myers-Briggs Type Indicator® (MBTI®) personality inventory is to make the theory of psychological types described by C. G. Jung understandable and useful in people's lives. The essence of the theory is that much seemingly random variation in the behavior is actually quite orderly and consistent, being due to basic differences in the ways individuals prefer to use their perception and judgment.
>
> Perception involves all the ways of becoming aware of things, people, happenings, or ideas. Judgment involves all the ways of coming to conclusions about what has been perceived. If people differ systematically in what they perceive and in how they reach conclusions, then it is only reasonable for them to differ correspondingly in their interests, reactions, values, motivations, and skills.

Myers-Briggs asks a range of written questions and identifies 16 distinctive personality types, including, for example, introvert vs. extrovert and thinking vs. feeling. And while millions of people have taken the Myers-Briggs inventory since it first appeared outside wartime use in 1962, it is criticized by some as being unnecessarily complex and difficult to interpret or even remember without notes. In 1991, a committee from the National Academy of Sciences concluded that some scales had high correlations to other similar tests, but a few scales did not (see Nowack, 1997). Even the test authors say that the accuracy of the inventory for an individual is based upon the person making honest responses: persons wishing to hide something unsavory in their personality might answer as they think they *should answer*, not how they actually feel. (See Quenk, 1999, or Pittenger, 1993, for dissenting views of the validity of the Myers-Briggs instrument.)

Despite these criticisms, Myers-Briggs is still widely used (for example, incoming plebes take the Myers Briggs almost as soon as they arrive at West Point), and even if not perfect, it still provides a springboard for understanding and discussing individual differences, decision making, and comprehending how one person would treat another, especially under stress.

A far simpler personality inventory has made its appearance in the past couple of decades, Real Colors (Figure 5.2), a copyrighted product protected for use and sale, as is Myers-Briggs. The Real Colors website (realcolors.org) says:

> NCTI's Real Colors® Personality Instrument is a leading edge tool that bridges temperament theory and real life applications in a way that is easy to understand, fun to learn and that offers unprecedented levels of retention. Using Real Colors®, people learn to recognize, accept and value the differences in others while improving understanding, empathy and communication.

Having gone through an actual exercise myself, the Real Colors inventory seemed simple to administer and simple to remember because a given personality type is reduced to a color (blue, orange, yellow, and green). However, reference to validity data for the inventory is not listed on the Real Colors website (although it may well exist). More importantly, the Real Colors inventory results explain not only what a person's values and personality type are based on the 10-item inventory but also how to get along with team workers or subordinates represented by opposing colors and how to make the best of that diversity.

Real Colors Personality Instrument®

	Blue	**Gold**	**Green**	**Orange**
Orientation	People, feelings	Accomplishments, tasks	Ideas, information	Action
Values	People, harmony	Dependability, hard work	Rational thought, curiosity, logic	Risk-taking, competition
Needs	Authentic relationships, care for others	Stability, order, structure	Independence, intellectual challenge	Physical activity, attention
Strengths	Empathetic, accepting	Organizational skills, detailed planning, follow-through	Problem solving, analyzing information, objective	Energetic, persuasive, leadership
Joys	Helping, harmony, romance	Traditional values, security, order	Discovery, understanding things	Trying new activities, competing (winning

Adapted from NCTI, *Real Colors*, 2005.

Figure 5.2 The Real Colors Personality Inventory is easily administered and offers a snapshot of a person's motivations, interests, and values.

The same criticism of the MBPI could be leveled at Real Colors, no doubt: no control groups, no randomized subject assignment to groups, and therefore not based on true experimental methods. But like the MBPI, Real Colors doesn't claim to be perfect—just an indicator of individual difference and discovering how a person undergoing the inventory might react under stress.

I asked some trained faculty to administer the Real Colors inventory as a class project. As promised, it took only about a half hour to administer and another half hour for assessment. But the out-of-class discussion about what we discovered about ourselves and our interactions with others kept going for the rest of the semester. It was quite a useful project.

A third values inventory available to an organization has been created by research conducted by Hogan and Hogan (see hoganassessments. com) and once again is a copyrighted product protected for use and sale by the parent company, which says, "The Motives, Values, Preferences Inventory (MVPI) is a personality inventory that reveals a person's core values, goals and interests. Results indicate which type of position, job and environment will be most motivating for the employee and when he/she will feel the most satisfied."

Organizations can use this information to ensure that a new hire's values are consistent with those of the organization. The MVPI can also help diagnose areas of compatibility and conflict among team members. The MVPI claims to identify core values that are part of a person's identity. Consequently, they are a person's key drivers—they are what a person desires and strives to attain.

The MVPI website says that in taking the inventory, a 15- to 20-minute exercise that has been validated on 100 companies, it identifies a person's core values and what motivates a given person taking this self-administered inventory.

Recognition—responsive to attention, approval, and praise
Power—desire for success, accomplishment, status, and control
Hedonism—orientation for fun, pleasure, and enjoyment
Altruistic—desire to help others and contribute to society
Affiliation—desire for and enjoyment of social interaction
Tradition—dedication, strong personal beliefs, and obligation
Security—need for predictability, structure, and order
Commerce—interest in money, profits, investment, and business opportunities
Aesthetics—need for self-expression, concern over look, feel, and design of work products
Science—quest for knowledge, research, technology, and data

The MVPI is somewhere between Myers-Briggs on complexity and ease in interpretation and Real Colors, which some say may be too simplified. Nevertheless, these three personality inventories will provide reasonably simple assessments of what is of vital importance to the test taker. They will vary on cost, the ability to self-administer, their availability online, and so forth.

Regardless of the inventory, all major theorists and academics from industry to the military will say that growing an individual's leadership potential begins with a candid self-assessment of his or her core values. We first understand what drives ourselves and then what drives others. These understandings of central values help us to meet in the middle.

We declare what we believe in and every individual who wants to lead subordinates, whether CEO, office manager, or department head, wants to move toward acting congruently with his or her core beliefs.

Now what? Let's assume that you or your group has undergone an inventory and has spawned a series of discussions and heart-to-heart talks among yourselves about what really matters in your life, whether you think people ought to be responsible for themselves in a safety or engineering context. Whether safety itself ought to be a fundamental core feature of what a leader and his or her organization does; whether that comes before anything else. Most of all, whether you really believe that safety and the individual come first, and whether you can behave in the same direction as those beliefs. If so, we've finally hit the road to authentic leadership.

Making a decision to behave congruent with one's central values: What triggers it?

It is often said that there is a triggering moment in a person's life when he or she sees and comprehends that fundamental truths about himself or herself have been revealed. You've heard an actor say on a movie somewhere as he is hanging off a cliff, "if I ever get out of this mess, I'm going to change my ways."

Almost every major safety leader I know says they had a triggering event in their lives when, after a time of reflection, they decided to become more, to begin to lead others in their world toward behavior that "walks the walk, and not just talks the talk." These revelations are never planned, but yet they happen, maybe at a conference, maybe on a road trip. I wish sometimes I had made a list of all of these stories and triggering events in my own life. But each and every one ended up with a personal and quite voluntary decision to reflect, to assess, and then to change his or her own behavior, to model the behavior he or she expected from his or her subordinates.

This is the very essence of incipient leadership.

Such a personal revelation happened recently in a conversation I had with a talented safety leader about a mining incident where people lost their lives. My colleague was called upon to help with the on-site investigation; he spoke with families who lost dads and brothers, and he began thinking about what really mattered in his own life. He told me he became introspective, and for a week or two after the investigation, he was sullen and isolated himself from others. Even without a personality inventory, merely being faced with life's realities sparked a deep and personal search for fundamental truths for my friend Josh. It affected him deeply and it showed.

I had known Josh for two decades when I heard his story. I had long wondered how and why Josh had become the guy that everyone in the industry looks up to—the guy who puts people's lives first in each and every thing he does or say. After finally hearing his story, I finally realized what initiated his commitment to safety today.

When I was still in college, a tragic event, triggered real introspection and a conscious decision to change the conditions that let the event happen.

As part of a three-person crane crew at a summer job, we had to trust each other implicitly and provide hand signals to each other and to the crane operator. We became very close after three months, but this careless organization did not providing fall protection (and we didn't know enough to ask for it), my buddy ended his summer job by falling backward into a two-foot diameter pipe 22 feet off the ground. He lived and graduated college in a wheelchair. It was my own "mountain top experience." I recall that after being in a funk for a while over my friend's terrible injuries, I made a conscious decision that I was going to work to the best of my own abilities to not let that sort of event repeat itself.

A sample Code of Eight Central Values is listed in the following, which I have gleaned from many sources. Until you are comfortable with understanding your own motivations, you can adopt these for yourself because they are an example of what many people find to be central to their own existence.

> *Self-reliance*: The notion that people are responsible first to themselves and will not ask for help until other avenues are exhausted.
> *Personal responsibility*: People are responsible for what they say, what they do, what safety actions they take or ignore, even how they spend their money.
> *Honor*: The act of carrying out, acting, and living the values of respect, duty, loyalty, and selfless service (taken from part of the U.S. Army Seven Core Values).
> *Loyalty*: Be dependable and respectful to your family, your spouse, your company, both in private and in public.

Respect: This is no more than another way of expressing the Christian doctrine of the Golden Rule: Treat others as you wish to be treated yourself.

Integrity: This means do the right thing even when it hurts. It means looking in the mirror and seeing a person who can smile about his or her actions that day.

Personal courage: This means not just physical courage, but knowing that leaders will sometimes have to risk their personal safety to help someone in need or to stand up for people who cannot defend themselves.

Safety: Every person on our site, in our offices, in our vehicles, and in our buildings deserves the highest degree of protection from loss of any kind.

An organization's core values, if they are sincere and genuine, will perfectly reflect the core values of the organization's own leaders. Once an individual's core values are identified, that leader can balance his or her strengths with a sometimes more diverse set of values in or among subordinates, the better to grasp all possible sides in a dispute or important decision. Hogan and Curphy (2008) point out that a leader with diverse but genuine values may initially provoke tension and conflict within the group, but this will also make it more likely that a broader variety of problems and solutions will be brought forward for discussion. To summarize before we move on, the literature suggests strongly that the path to becoming a leader require a choice to become a leader based on a life-changing event, and an honest assessment of a person's core values.

chapter six

Culture, safety, and engineering

Earlier, we encountered West Point's Col. Bernie Banks talking about the imperative to teach leadership in the 21st century because of the ambiguity and fluidity of the challenges in today's global environment. He also endorses the notion that an authentic leader is one whose behavior is congruent to his or her central, core values.

Banks is a proponent of studying Edgar Schein's (2004) *Organizational Values and Leadership,* suggesting that there are ways to understand organizational culture and that there certainly ways to influence it in the event that the organization's behavior, its culture, is not consistent with its policies (Figure 6.1).

The term *culture* has been around for a long time and parallels the popular use of the word *paradigm,* coined first by Thomas Kuhn in the 1940s and made popular in a big way in his 1968 book, *The Structure of Scientific Revolutions.* Everybody says "culture" and everybody says "paradigm" whether or not they know anything about them.

Like the idea of culture, the use of the notion of paradigm was at first usually restricted to scientific inquiry in the physical sciences, chemistry, and physics. But as time went on, the notion of paradigm became less restrictive and popularized in books by Joel Barker and others (for example, see *Paradigms: the Business of Discovering the Future,* 1993). These writers brought usage of the term *paradigm* into corporate offices. Maybe the word has lost some of its original punch because it is overused now.

The word *culture* is similar in that it was originally the stuff of anthropology, and now, it is the stuff of boardrooms. The first usage I can find placing safety and culture in the context is an analysis of incidents in a car manufacturing facility in 1951, where the "safety climate" was an issue (Keenan, Kerr, and Sherman, 1951). Popular use of *culture* in safety took off in the 1990s, and even OSHA uses it today. Still, scholars turn first to Edgar Schein and then to Gerte Hofstede, who we'll meet later.

Certainly, the popularization of the idea that organizations have a culture is due to Dr. Edgar Schein, who is still a professor of management at the Massachusetts Institute of Technology. His seminal 1992 book, reprinted in 2004, is based on his years—decades, really—of experience with real companies he has worked with. In fact, he is quite and justifiably proud that his theoretical work comes directly from quasi-experimental research experience with real people in real organizations.

Figure 6.1 Col. Bernie Banks, PhD, is a chopper pilot and has five master's degrees. In 2012, he became chair of the Department of Behavioral Science and Leadership at West Point.

Schein (2004) defines culture as "a pattern of shared basic assumptions learned by a group as it solved its problems of external adaptation and internal integration, which has worked well enough to be considered valid, and, therefore, to be taught to new members as the correct way to perceive, think, and feel in relation to those problems."

Schein applies the anthropological notion of culture to organizations and suggests that organizational culture emerges when groups attempt to solve problems. When they do that successfully, patterns emerge. "Organizational cultures, like other cultures, develop as groups of people struggle to make sense of, and cope with their worlds" (Trice and Breyer, 1993, in Schein, 2004).

Our world has groups organized and working toward a common goal—baseball teams and university faculty, logging companies, and governments—and as they solve their own particular problems, all will develop an organizational culture. Whether it is getting men on base in the ninth inning or getting promoted and tenured, sawing trees efficiently into cants, or protecting citizens overseas, these groups work in ways that can be described as a culture.

But Schein says that when leaders cannot understand or predict the behavior of their own organizations' cultures, new policies and even new great leaps are likely to fail because they are inconsistent with the organization's culture. Dan Peterson says this in almost the same words in his last book, *Techniques of Safety Management: A Systems Approach*, published in 2003. As 1 of his 10 guiding principles, he says, "A safety system should fit the culture of the organization," and suggested that mismatched programs and cultures are certain to fail.

Schein dismisses the idea that culture is merely an implicit, hidden aspect of social life that ends up as the glue for keeping groups together, and largely unknowable. Rather, he sees organizational culture as an explicit expression of behavior and therefore knowable, measureable, and predictable. This concept is much more useful in our study of leadership because if we can measure behavior through its appearance and expression, we can predict what will happen when those variables change. Schein says in his 2004 book, "Any group with a stable membership and history of shared learning will have developed some level of culture." Let's look further at how Schein classifies levels of culture so we can grasp a greater understanding of how it can be used, and if culture is somehow mutable, how it can be used to actively transform the organization.

The three levels of an organization's culture are represented in Figure 6.2, widely used in management and leadership theory courses and taken from Schein's 1992 work.

Schein's research suggests that there are three levels with which to evaluate an organization's adherence to organizational core values.

An authentic leader is one who displays levels 1, 2, and 3 in the figure. That is, an authentic leader will have artifactual values and stated values as long as he or she also displays actual values at the same time. These values show up in the organization's culture.

According to Schein, observed behavior, job descriptions, and how the organization is supposed to work are all artifacts of the culture, easy to observe but harder to decode. Artifact culture is what a new hire sees on the first day of work: office space allocation, who has the most modern technology on his or her desk, who seems to be well dressed every day, and so forth. All of these are easy to observe but difficult to decipher.

Schein offers a perfect example of how artifactual culture is not enough to judge a culture and is sometimes confusing. He observes that pyramids, to the Egyptians, were widely used tombs, but culturally,

The three levels of values

Level	Values	Appearances	Authentic
Level 1	Artifactual values: what people *say* they value or *how* values appear (you *see* that aspect of culture but it might be all for show and no real action)	Example: company logo with a green cross embedded in the logo	Not clearly congruent with actual values but could still be authentic
Level 2	Stated values: espoused values (you *hear* that aspect of culture but it might be all talk and still no action)	Example: policy stating "no tolerance for drug use at work"	Not clearly congruent with actual values but could still be authentic
Level 3	Actual values (you *live* that aspect of culture; this person "walks the walk and talks the talk")	Example: stopping an unsafe act or condition without being told to do so	Congruent and authentic leadership

Figure 6.2 Authentic leaders are characterized not by just what they show or say but also what they do.

pyramids—structurally the same arrangement of stones—meant something very different to the Mayans, who used them as temples. These are public and observable examples of a culture, but the meaning on its face is difficult to decode. These symbols are ambiguous because there is nobody there to translate their meaning, particularly separated in time by millennia. We only think we know what the artifacts mean because we are projecting on them our own biases. When we judge culture from artifactual examples only, we are guessing at meaning.

In fact, Schein suggests that it is probably dangerous to interpret culture and its meaning from only its artifacts. His brilliant comparison of loosely or tightly managed organizations (*loose management* suggesting lax management oversight or inefficiency and *tight management* suggesting lack of innovation) shows expressions of what we are projecting on the artifacts of a culture. There is no translator there to give us a better representation of what the overt expression of a company's culture might really be. We have to interpret this expression of culture for ourselves.

Look at Google, the huge organization, where people come to work in jeans, have lunch and coffee provided free, and are encouraged to take naps during the day. Without a translator to tell us what this loose organizational culture means, we'd miss that Google is one of the best led companies in the United States, with a strong financial portfolio and a waiting list of thousands of engineering and computer science job applicants.

We have to be careful interpreting culture only on artifact. If we're going to try to change a company's basic cultural fabric to value safety more than it does, for example, we'd be smart to look at, but not pay too much attention to, the company's artifactual culture.

Using the same framework, the next level of culture is espoused values and assumptions, but unfortunately, we still don't have much to grasp when considering culture change here either because these espoused beliefs, beneath the surface, can be just words and talk, aspirations even, and can even be contradictory to what the company does. What if a company's mission is "to preserve and protect the people, property, and efficacy of the organization" but the company's Chief Executive Officer (CEO) refuses to wear glasses and hard hats on a walk through? Obviously, there is a disconnect between the espoused values to preserve/protect and the observed behavior of ignoring basic safety precautions on the construction site. The company's espoused values are not congruent with its behavior, yet if the company top brass followed the posted rules and embraced the safety policies throughout, we could conclude that there is congruency between behavior and espoused values.

There is no point in trying to "unfreeze culture, change culture, and refreeze culture" (an analogy attributed to Kurt Lewin, the father of modern sociology) when we don't even know exactly what culture we are unfreezing.

How do we find out what a company *really* values in order that we can think about changing its basic culture toward something more adaptive or toward a total-safety culture where everybody is really involved? We must go deeper still into the core value stream of the company, to its underlying assumptions. These are the actual values-in-use, according to Ott (1989), or values-in-actual practice. These are so strongly held that they are difficult to give up and maybe impossible to give up unless the employee simply leaves the organization.

Once basic assumptions are so ingrained in a company that they become second nature and not consciously considered, they are taken for granted. If a company has "safety first" posters on its walls, it suggests values of protecting its workers as its highest objective. But as a matter of course—without thinking—if the same allows workers to operate the controls of a second-story grain elevator near the edge of the roof, and without fall protection, then the company's values-in-practice are incongruent to both artifact culture and espoused culture. The company says that it values employee safety but it makes safety optional in practice. This is the sad mismatch of "what we say is not what we do"; everybody knows it, everybody does it. The behavior of accepting unsafe acts is just a norm held by the group when there is incongruence between levels 1, 2, and 3 values. And leading people in a culture where there is a disconnect between values-in-use and artifactual/espoused values may be distasteful or impossible.

Schein says that people see and know the incongruity, and that it causes stress, which is dealt with by distorting the actual truth or "falsifying what may be going on around us." We may ignore the values-in-use mismatch. When a serious fall occurs, it is chalked up to bad luck; after all, look at our mission, the posters in the break room, and the safety stickers we hand out every month. Imagine the pressure on a new craft worker who sees the obvious: workers on elevated surfaces walking next to the edges of an unprotected roof. If he or she is absorbed into the organizational culture, he or she may call attention to the unsafe act but may get fired in the process. If he or she is complicit, the organizational culture has done its job of normalizing and flattening those who are idiosyncratic (Figure 6.3). Either way, it is stressful when there is a culture of mismatch between organizational and individual values.

Back to those of us who have waded carefully through our company's artifactual, espoused, and actual values only to find no incongruence. Congratulations, now we can begin to think about the "unfreeze culture, move the culture toward even more individual responsibility, integrity and so forth, and then, refreeze culture" process.

Hannah says the same thing in Crandall's (2007) text that a leader who is self-aware and authentic in his or her own right can actually change an organization where those virtues are also demonstrated.

I have argued that authenticity is itself a value held by the leader; thus, inasmuch as leaders can get their followers to emulate them and internalize those same values, they can diffuse shared values of authenticity in the culture (of the organization). In addition, if the leader manages the authentic development process of followers through dedicated goal setting and triggers experiences and ion periods of reflection, over time, they can raise the average level of authenticity across organizational members.

This simple strategy for building culture from the stepwise progression I am describing here (self-awareness to core values, core values to authentic leadership, and authentic leadership to authentic culture) is depicted in Figure 6.4.

So before we think about changing an organizational culture to attain an authentic, values-driven culture, we have to examine our own values and codify them; then we exemplify those values and ensure congruency in our own artifactual, espoused, and actual values; and then we're ready to raise authenticity in our company.

There is a great quote attributed to Ralph Waldo Emerson about values incongruity:

"What you do speaks so loudly that I cannot hear what you say."

Figure 6.3 The irony is lost on few people when a mismatch exists between espoused and values-in-use.

Self-awareness leads to establishing
personal core values

Exemplifying core values leads to
authentic leadership

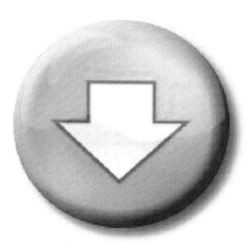

Authentic leadership leads to authentic culture

Figure 6.4 A values-based culture rests on the beliefs and values of its individual members.

Authentic leaders can exist at all levels of an organization; a forklift driver does not need the title "CEO" or "foreman" to manifest his actual values in voluntarily instructing a new employee in the virtues of inspecting the truck's brakes every day whether the rules require it or not. When you think a behavior is manifested because it's the right thing to do, it reflects authentic leadership.

My son, Austin, is now a piping engineer for an international corporation. Working one summer for another company while he was still in college, he described a perfect example of an authentic leader who worked in an inauthentic culture. The rig manager, a young drilling engineer in his own right, expressed to my son that he was aware of the crippling possibility of heat stress with his men working in fire-retardant full coveralls and personal protective equipment (PPE) in that summer of 2010. The rig manager brought in two huge 10-year-old fans for the dozen drill hands in an effort to keep his subordinates cool under the stifling conditions. The rig manager had stored the fans at his farm to prevent theft and was happy to give the fans a new and useful lease on life.

After only a few hours in the first day of their use, some official inspector showed up at the drill site and cited the manager for non-OSHA compliant use of fans since the label had worn off and it wasn't clear that they were still compliant with some code or other. The manager obediently removed the fans and then after work and on his own time, drove three hours one way to purchase two more fans at a late-night supplier. He ended up driving almost all night and brought back the brand new fans, this time fully labeled and ready for the 6:00 a.m. start of the shift.

The rig manager didn't brag about it; he didn't even tell anyone about his all-night drive. My son noticed that the fans were brand new and asked about them; otherwise, we'd probably never know this story. The rig manager's behavior was fully consistent with his concern for his crew. That is authentic leadership. That kind of leadership is as good as it gets.

In Crandall's (2007) book *Leadership Lessons from West Point*, Sean Hannah reflects somewhat tongue in cheek about "spotlight Rangers." These are young Army Rangers-in-training who do and say exactly the right thing when the instructor is around but who act irregularly otherwise. The spotlight Ranger represents true Ranger values only when the spotlight is on him, but as Hannah says, he is soon found out, rejected through peer evaluations, and washed out of Ranger school.

Hannah notes:

> [Authentic leaders] are highly aware of social cues and followers' needs, expectations, and desires. This awareness allows them to react to their environment and make certain aspects of their true self more salient than others at any time. What is critical

here is that they bring to any situation part of their true self but not a false self. This nimbleness results in what psychologists term a working concept that is adaptive and responsive to situational cues and is situation specific, yet is a subset of their true self.

Authentic leaders, Hannah goes on, are pretty much always under some sort of scrutiny, and things will quickly "come crashing down should an [authentic] leader lapse or be uncovered as pseudo-authentic." The once authentic leader has a more difficult time recovering.

The take-away message here is this: An organization's culture is dictated by the values held by its leadership and displayed publically as artifactual, espoused, and values-in-use. The culture can't be bought or copied from a book somewhere. If the organization is truly and authentically values based, its actions must routinely and perpetually be dictated by those same values embraced by its leadership. If "safety of employees" or "respect for each individual" is a core value, then the company's actions, words, and daily work are always held in the white light of scrutiny.

chapter seven

How we can change organizational values and why it's important

I have discussed the crucial importance of assessing core values and the use of three of the many inventories that are available on the market. Any of these gives a snapshot of individual values. Understanding what the organization's people feel at their root core is a necessary step before moving to any consideration of organizational values. When we understand our own motivations, we can work to understand and build bridges to others with different core values.

From there, we discussed Schein and Hannah as they treated assessing culture and the importance of congruity among the various levels where values are displayed. Now, let's examine ways you can help others to internalize personal and organizational values, that is, to make them values-in-use.

I especially like Chip Daniels' treatment (Crandall, 2007) of methods that can be used to help individuals and subordinates to internalize organizational values once they are established. Daniels' treatment is easy to understand and easy to take back for discussions among subordinates.

His five steps to internalizing values among subordinates and, ultimately, the whole organization are these:

1. Hire the right people to begin with, those who *self-identify and self-select*: This one is pretty easy to understand. Find people you want to work for you who have already expressed values or hold a set of values dear. That would be a Girl Scout with a Gold Award or a Boy Scout with an Eagle Scout Award. When I interview these people for graduate school or to work on my research projects, I can tell a mile away about their scouting past, and many of these students can recite the "trustworthy, helpful, courteous, kind..." by heart. These people will self-identify with a strong set of values that they normally follow in life. Of course, scouting is only one example.

   Because these people have self-identified as having core values, these individuals should also be among the first to assume leadership roles themselves.

This set of people would probably also include former enlisted or officers in the military, or who were on the Reserve Officer Training Corps (ROTC). Similarly, farm kids have worked hard toward common family goals, and are usually willing to work hard and not whine about it.

I recognize that my examples are regional and that they may not apply to you, but they are not trite or irrelevant. In my world, I have hired sons and daughters of coal miners and also farm kids of both genders for my research projects and I have never been wrong: the groups have a subtle but real set of values based on hard work and self-reliance, and those values are infused into the families. Strong family orientation seems to bring a strong measure of self-worth, and loyalty, regardless of region or geography.

If you are working now, try my method of looking for farm kids, Eagle Scouts, and sons or daughters of coal miners or any other labor-intensive industry when you are hiring a new safety professional or project engineer. In 25 years, I have never once gone wrong with this method.

You don't have to teach these your people how to say "yes, sir." They already do it. The point is to look for applicants with core values congruent with the organization's core values. They will instinctively infuse their own values into the new environment.

2. Create an *early socialization process*: Values education is what this point is about. If the organization has a fully developed set of core values at the ingrained and involuntary level (not just artifactual or espoused), employees should be instructed in them and periodically reminded of them. Socialization can be done during new-hire orientation, during just about any training, and at opportunities in between, that is, "This is what we stand for." If a new hire is assigned to a mentor, the mentor should explain the organization's core values.

   Parsons, URS, and Jacobs are companies I am very familiar with because they have hired dozens of my graduates over the years. They have deeply seated core values, and these huge corporations make a great effort to educate their own new hires about corporate values. Company literature reminds current employees that "this is what we stand for, and this is what we expect from employees." This is early and regular socialization.

   Employee families can be enlisted to share in the socialization. Daniels says, "If the employee shares the organizational values but the spouse does not, stress will eventually come between them." The effect of inculcating a strong set of safety-related core values into families is sure to have positive rippling effects: If mom and dad believe in a duty to make sure their at-work subordinates are prepared with the right PPE, they'll probably take that message home to the family, whose members will be inclined to adopt it.

I know this isn't necessarily true of all family farms, but at our family farm, we have always had rules about using crane operator-type hand signals for stop and go on the tractor, and the stop signal always means *stop now*. After doing this with my family for two decades, and also with every kid who helps with family hay baling, we now have two generations of young people helping us each year in our hay field that have adopted and use these unambiguous safety signals. Ditto the use of hearing protection around machinery; ditto staying far back from power transmission shafts. Early socialization works at the small scale, too, even the family farm.

3. *Use of role models*: I have a lot of respect for Kiewit, a huge, family-owned, multinational construction and mining company based in Omaha, Nebraska, and employer of about a dozen of our graduates over the years. The thing I admire about Kiewit hiring our graduates is that for a year—a solid year—these new hires are accompanied by a sponsor, a role model, a mentor, who shows the new safety professional or new engineer how to act and how to exemplify the company's core values about safety. These mentors *live safety* and they *mean safety*. The Kiewit mentorship program is one reason the company's core values are congruent to their artifactual, espoused, and actual, base-level, ingrained values. Kiewit shows it, acts, and does it.

When recommending the use of role models to exemplify a company's core values, I can't say it better than Daniels himself does: "Role models reinforce proper values and try to help their protégés make sense of what is happening in their lives. This monitoring relationship and investment in protégés is crucial to long term satisfaction."

In my own simple way at the university, I recognized early that I could also be a role model, and I am quite conscious of the fact. In my humble position as a faculty adviser, I can pass along a few truisms that might affect their lives. For example, I tell students that they have only a few years here at school to shape their futures as graduate safety students and engineers and that their clock is running. I tell them to find themselves a hero with real values and emulate that person, whoever he or she is. Sometimes, I tell them to go sit on a rock someplace and discover their own deepest values and then follow them. I tell them to plan their own luck because nobody else will.

Using simple techniques, I have tried to pass along wisdom and research findings about safety and engineering, and as a result, I have been lucky enough to transform lives. In this role as a listener, an adviser, and perhaps, a role model, I am watching these young people take on the world. I admit that am the luckiest person ever, and *you can be that person, too.*

4. *Sharing of stories and examples*: As Daniels says about values-consistent peers, "supervisors… and even direct reports [employees reporting directly to you] can serve as powerful role models that heavily influence others in the organization." Dinner or casual conversations with family or subordinates can be about which person made the right— or wrong—choice. These stories are not meant necessarily to be moralizing or cast the teller in a positive light as much as they are used to convey a message that this "values-consistent" person did something worth copying, or, if the action was inconsistent, worth avoiding.

Classrooms, break rooms, locker rooms, and board rooms are equally useful places to tell stories about a person who removed the guard on the table saw, experienced a near-miss, and then replaced the guard. Maybe it's a story about the person who routinely does roof work using fall protection after learning the hard way in a close call. Real stories are the best.

The following advice is based only on my own anecdotal research. I have amalgamated in a five-part guide for storytelling. Stories that should be told are

1. About an actual close call, a real case with familiar people.
2. By a person with credibility; whether it is craft worker or manager is not important.
3. Most often told about a person who discovered the right thing even if he or she didn't actually do it.
4. Not directly in response to an *injury* that just happened. Near-miss investigations do not fall in this category. A story following an injury would look retributive.
5. Not condescending or heavy handed.

Here are three sample stories you can use:

Telling the story of the Vietnam-era pilot, Hugh Thompson, always makes a heavy impact about the success of values-in-use when listeners realize this guy took his life in his hands at the Mai Lai massacre site. Thompson landed his own helicopter between out-of-control GIs and protected civilians because it was the right thing to do. Thompson intervened personally, showing values-in-use. It makes a great story about doing the right thing.

I related in the previous chapter a story about my college buddy who fell backward into a vertical concrete pipe makes a similar impact about failure of a safety supervisor's values-in-use. This poor guy, my friend, was supposed to walk out on the 22-foot-tall pipes standing on end—wet with rain and coated with grease to prevent concrete from sticking to the steel end-ring. My buddy made it to the end of summer before he fell backward into a pipe and broke a dozen bones, including his spine. The company had lots of safety posters, but there were no safety values-in-use that day. As I said, he graduated from college in a wheelchair.

Daniels relates the story of Johnson and Johnson's chairman, James Burke, and his "masterful handling" of the Tylenol cyanide tampering. Back in 1982, he says, the public was panicked by a spate of poisonings that happened due to package tampering. Burke recalled all Tylenol capsules and offered to replace them with tablets; he put out a media campaign to reassure the public and he immediately started packaging Tylenol in the familiar tamper-proof foil seal, which other pharmaceutical manufacturers have copied almost universally. Nobody required him to do this, and he did it all before the predictable wave of public criticism could descend. Burke displayed values-in-use.

What we are trying to do through storytelling is to teach about personal responsibility, or a duty to intervene, or service to the community. These kinds of stories are remembered a lot longer than preaching or some silly, plastic values card, and they are easy to retell. The idea is to make values-in-use automatic. When safety seems corny or out-of-date, a real story with a real impact is usually just around the corner.

5. *Use of feedback and performance evaluation*: Some organizations use formal, written reports to evaluate omission or commission of the organization's expressed values. The importance of the written evaluation is this: It tells the organization that the employee is, or is not, upholding its values, that there was a values infraction (lying or cheating the system), or conversely, the employee is recognized for doing the right thing. The values performance, says Daniels, is just as important, maybe more important, than the employee's basic job performance. It predicts well into the future, for one thing, and better than mere adherence to what the job requires.

If we summarize what we have learned from the Schein readings, we can say that it is important that a leader transforms a culture—even one in a department or small work group—into one supporting congruency with organizational values. It has been said that Schein's main contribution is the admonition to leaders to study and understand their own cultures because strategies and plans are likely to fail in the event the leader's vision does not match the culture, or when safety consists of just posters and catchy slogans.

Another researcher examines culture at the international level when he suggests that a leader's vision must match not just the organization's culture but also the culture inside a particular country or region of the world; conflict is sure to arise when they are not congruent. Dr. Geert Hofstede started out in mechanical engineering. He switched fields after years of work at IBM and became a research psychologist. During both of his terms at IBM, he analyzed data on over 100,000 individuals at IBM who worked in 40 countries around the world.

His research, highlighted in his 1994 book (and updated and reprinted in 2010), *Cultures and Organizations: Software of the Mind*, suggests that there are national, even regional, groupings of culture along some fairly persistent dimensions. Why should we care about that? Because, as he says, "Culture is more often a source of conflict than of synergy. [Ignoring] cultural differences is a nuisance at best, and often a disaster." This means that a company's upper management wanting to plan a new operation in an entirely new region of the country or the world had better pay attention to what is important to the locals.

I have always valued the century-plus, global perspective of *The Economist*, and so I was glad to be able to use an article about Hofstede I had squirreled away for future use. The November 28, 2008, issue of *The Economist* discussed how leaders who work across international boundaries have special sets of challenges. The article cogently summarized Geert's four original cultural dimensions, which I cite in the following verbatim, along with some salient conclusions drawn by the author (Hindle, 2008) in *The Economist* article:

- Individual versus collective (IDV). This refers to the extent to which individuals in a different country might naturally expect only to look after themselves and their immediate families. This is sometimes reflected in the use of words such as "I" and "we," "my" and "our" more than we are comfortable with in the US or Western Europe.
- Power distance index (PDI). This refers to the extent to which a society accepts that power in institutions and organizations is distributed unequally. Countries where the PDI is low generally favor decentralized organizations, whereas those with a high level of PDI are more accepting of centralized authority.
- Uncertainty avoidance index (UAI). This is the extent to which employees feel threatened by ambiguity, and the relative importance that they attach to rules, long-term employment and steady progression up a well-defined career ladder.
- Masculinity (MAS). This refers to the nature of the dominant values in the organization. For example, is it predominantly influenced by masculine values such as assertiveness and monetary focus, rather than feminine values such as concern for others and the quality of relationships?

Hofstede subsequently added a fifth dimension after carrying out a study of Chinese managers and workers during his time in Hong Kong. This he called long-term orientation (LTO), which refers to the different time frames used by different people and organizations. Those with a short-term view are more inclined towards consumption and to maintaining face by keeping up with the neighbors. With a long-term attitude, the focus is on preserving status-based relationships and thrift.

Hofstede developed a system for scoring individual countries according to their culture. The differences can be dramatic and surprising. Greece, for instance, scores 11 on the UAI dimension, while Denmark, a fellow member of the European Union, scores only 23. Less surprisingly perhaps, Sweden scores only 5 on the MAS of its organizations, while persistently chauvinistic Japan scores 95. On LTO, while China excels with a score of 11, the not-so-far-away Philippines scores a mere 19 (Hofstede, Hofstede, and Minkov, 2010). They are different and we need to pay attention of we expect to collaborate there.

What does this mean to us? It simply means that what values and beliefs are held in one region may not be held in another country, even another state or another city where the company wants to build a new plant. The careful leader seeking to espouse and standardize the company's central values will assess them to avoid conflicts at first and downright derision later. For example, GE was always careful to assess its local culture, even globally, and for the most part made sure to synchronize its core values with its public. On the other hand, the U.S. Army got it wrong when it developed a "values dog tag" to go with a soldier's dog tag (personal identification and emergency information) around his neck. The values dog tag did not fit the culture because it was not authentic—it was merely handed out and required that it be worn. Hardly surprising, it was not well received by regular soldiers, as Kolditz reports in Crandall (2007):

> There is, however, a fundamental flaw to these approaches [forcing soldiers to display values in simplistic way]: they lack authenticity. I recall watching several military formations and groups receive the "new" values. When the plastic dog tags were issued to soldiers, the event was often marked with howls of cynical laughter, and the cynicism was most apparent among soldiers with the most combat time and skill.

A leader who forces values on its subordinates will, as Kolditz relates, experience push-back and outright hostility. The values have to buddle up and be embraced in their own time.

Hofstede and Kolditz warn of the importance of getting it right the first time by assessing regional and local belief and values systems because you only get one chance.

What happens when there is resistance to change? Introducing the James–Lange theory

So far, it seems that the answer to the question posed in the chapter title ("Can we develop organizational values?") seems to be yes, we can. But what happens when there is resistance to change, either from the old guard, who are intimidated by new ideas such as establishing an honor code, or those who just don't care?

There is some counterintuitive research that says we can apply pressure to set up change in organizational norms. Let's take a look at the James–Lange theory of attitude change. In brief, it says that *acting the part* eventually leads to acceptance.

The *Corsini Encyclopedia of Psychology* (Corsini et al., 2010) is a four-volume set of authoritative research by over 600 individual psychologists. The *Corsini Encyclopedia* describes an emotion as "a strong mental state usually accompanied by excitement or high energy that gives rise to feelings and passions" (p. 495) (Figure 7.1).

William James and Carl Lange, early American psychologists who still have immense impact on the field of psychology and human motivation, came to the same conclusions about how we might cause people to become receptive—even emotional—about having a strong values system that acts as a yardstick for our behaviors. When we act in accordance with the values system (values consistent), whether we believe in them at first or not, the values grow stronger.

One very old theory of emotion suggests that emotion is first the result of physiological arousal, and then we interpret the arousal in our own individual ways with emotion as the result. If we do not interpret the arousal, we do not have the emotion.

> Under the James–Lange Theory, a leader first encourages, even requires, the behavior that is recognized as consistent with the values system. Eventually, when we act in accordance with the values system, whether we believe in them or not, the values actually grow stronger. Finally, they become automatic and cultural.

Figure 7.1 The James–Lange theory suggests that culture can be manipulated by paying attention to motivations.

The same chapter in the *Corsini Encyclopedia* continues:

> The James–Lange theory of emotion basically pos-
> tulates that emotions are made up of bodily changes
> (eg., arousal) and a mental event or feeling.
>
> An event was perceived, physiological changes
> occurred as a result of this event, and the feeling
> that one had as a result of the physiological change
> was the emotion.
>
> Sad events led to bodily changes that led to
> sorrow, while frightening events led to a different
> type of bodily change that led to fear. Most people
> believed that laughter was the result of being happy,
> and crying was attributable to sadness. James and
> Lange argued the converse: laughter [the bodily
> state] led to happiness [an interpretation of the
> bodily state, or emotion] and trembling gave rise to
> fear. (p. 495)

The James–Lange theory of emotion was counterintuitive a hundred years ago, and it still is today, but it is useful. Dr. Bernie Banks, whom we have met earlier, says that the James–Lange theory is at work when attitude change (an emotional investment in a values system, for example) is needed. A leader encourages, even requires, the behavior that is recognized as consistent with the values system (this part is important). The leader offers encouragement and training as necessary but holds the individuals accountable for acting consistent with the values system, even if they aren't aware they are doing it or don't care one way or the other. The behavior comes first, says Banks. Eventually, with care and nurturing from the leader (role modeling and storytelling—strategies we saw earlier in values adoption), the attitude follows and finally the emotional state and the values-consistent behavior.

Maybe that would have worked with the Army's values dog tag, but officers forced them onto their subordinates with little or no explanation. There was no storytelling, no modeling, nothing. No wonder there was resistance.

In the next chapter, we'll examine what a motivated individual can do to have his or her department, or even small group, act on the basis of fundamental core values without a day-to-day supportive atmosphere (what I call a "depleted" environment). As we'll see, culture change can occur even at the micro level when upper management doesn't care. After all, who is going to argue with a leader asking his or her subordinates to become persons of character and integrity?

chapter eight

A values-based leadership model for use in depleted environments

In Chapter 1, I pointed out that Casey Brower and Tom Meriwether at VMI both warned me that leader development is difficult without a fabric of support or, as West Point's Banks called it, an ecosystem where the larger organization from the top down has bought first into awareness, and then to expression, its *central values*. But because this topic has been talked about for a long time now, many organizations have developed statements about core values because they had decades to establish them.

What about leader development in a unit as small as a department or where upper management isn't supportive?

As I have talked with my students and working alums about small and medium companies that they work for, I have found that a streamlined version of "values-based leadership" is indeed possible even in these "depleted environments" where leader development is not a priority. A motivated and authentic leader can still have influence, and maybe even great influence.

I propose here a boiled down version of what is possible in leader development for a committed department head or craft workers who wants to change things. I call my three-part, simplified approach "Simplified Values-Based Leader Development in a Depleted Environment." I have borrowed heavily from people who discuss how to develop leaders in a restricted and highly controlled environment such as the military. Indeed, Hesselbein and Shinseki, whom we have encountered before, discuss developing military leaders, but what about industry, construction, even a school system or a small government operation? Nobody I could find actually talks about developing leaders in microcultures that do not support it. Yet, I think it is possible to create "leadership teaching moments" and to create a microculture of leader development even at the department level. Examples of teaching moments in microcultures abound if we look for them.

As an example of a simple teaching moment, one of my graduate students spent a summer in Abu Dhabi working for a major construction general contracting company, built mostly independently on a high ride building. Ryan B. told this story that perfectly exemplified the values-in-use qualities of a leader and, particularly, personal courage at the microculture level. He said he and another intern risked their internship by confronting the representative of a huge, big-name contractor when he allowed his employees to ignore fall protection on the 10th floor of their unfinished building. Behind their backs, the contractor told his employees to just ignore the interns and go about their work without fall protection after they leave.

Rather than cave in or be intimidated, and at the risk of losing their jobs if the big contractor complained to the high profile general contractor about these trifling, meddling interns, the interns stood their ground. They knew how devastating a fall from heights can be, and they knew the rules.

The interns returned with determination and faced down the contractor. Being diplomatic but firm, they explained the rules and the direct and indirect effects of a fall, and they explained the financial consequences to their contract, as well.

And they made sure the contractor knew they were coming back to make sure he was following the rules.

A few days later, the contractor's employees were using the prescribed gear. The interns, in an unsupportive microsystem, displayed personal courage and values consistency; they knew compliance's technical requirements and their own policies and they influenced a decision by forcing a test. Nobody is prouder of these young people than I am. They did the right thing even when nobody was looking, even at the risk of being fired. It's a story that can be told time and time again.

Establish an honor code

My SVBL model is for employees who are, or want to be, change agents at the supervisor or foreman level. Maybe the employee is simply an influential craft worker who wants his or her team to be the best it can be. He or she knows people aren't following the rules and there is a growing cynicism of policies generally. Is leader development possible in this tough environment? Tough, yes, but possible using the right approach.

In my simplified model, the craft-level leader tells his or her direct reports about what is expected, about what behavior the junior leader wants to see *at its most fundamental level*. Once a junior leader decides it's time to make some important changes toward values-in-use, the model is easily understood and used.

In part 1, I suggest starting with an *honor code*. An honor code is one type of accountability and feedback tool that says, "this is how we act here." Lacking a larger ecosystem of leader development that few small companies can afford, an honor code is one way to publically state its important values for others to emulate. One example of an honor code is used at VMI and the U.S. Military Academy at West Point (Figure 8.1).

Figure 8.1 Thirteen words: Many top industry executives say that their success has been made possible, in part, by adhering to the West Point Honor Code, which they learned when they were cadets.

This honor code is as simple as it can possibly be—only 13 words. Other schools have similar codes, including The College of William and Mary, the nation's oldest college (founded in 1693) and with the nation's oldest honor code even before West Point or VMI. Alumni from these and many other schools will tell you that even years after graduating, the honor code made a lasting impression on their lives and what they expect from others in their department or small group.

I have even used the same honor code in my graduate classes. At first, my students think this is trivial or beneath them, and certainly corny, but when I explain the context—that they will soon have their own subordinates for whom they will expect honesty and consideration of others—the room gets pretty quiet for a while. Apparently, nobody has ever asked them to subscribe to an honor code before. In a week or two, a student here and there will stop by my office and tell me quite candidly that they want to follow these high aspirations and pass them along after they graduate.

Is there a need for an honor code in a company without a complete ecosystem to support values-in-use? I say "yes" because safety leaders and engineers can never, ever, take employee trust and the public safety lightly. When people can die on your watch, and when there are no do-overs, our future leaders have to get it right the first time. In turn, employees have to be trusted to do the same thing. Even when nobody is looking.

Especially when nobody is looking.

When a leader has to be right the first time, when he or she is in charge of subordinates whose lives may be in peril, when direct reports have to do the right thing when nobody is looking, *then trust and integrity are more important than training.* A person's word must be an indelible bond.

Newton Baker was from Martinsburg, West Virginia, and he was President Woodrow Wilson's Secretary of War. He stated it clearly when he endorsed for the first time the notion of an honor code for the entire U.S. military because an employee or soldier who is in charge of others cannot trifle with the lives of his subordinates. Baker said:

> Men may be inexact or even untruthful in ordinary matters and suffer as a consequence only the disesteem of their associates or the inconvenience of unfavorable litigation, but the inexact or untruthful soldier trifles with the lives of his fellow men and with the honor of his government, and it is therefore no matter of pride but rather a stern disciplinary necessity that makes West Point require of her students a character for trustworthiness that knows no evasions. (Recounted by M.G. Maxwell Taylor, superintendent at West Point, 1945)

A dishonest worker may say that the electric power is turned off before telling another worker that she can change a breaker in a 480 volt service panel. A dishonest supervisor can say he has performed a walk through of morning conditions on the construction site and not bother to inspect the new utility excavation, allowing a forklift to back into it. When people can be injured or even die on your watch, there needs to be unassailable trust at every level, without asking, with nobody looking. Your subordinates must trust that you do what you say, and vice versa.

Having an honor code is a public expression of the desire to try to do the right thing at all times, under all conditions, and not tolerating those who will equivocate. This 13-word code can be easily adapted to use in organizational materials, into new-hire orientation, and it can easily be knitted into the real stories like those about Hugh Thompson, James Burke, and my college buddy who fell backwards into the pipe. I recommend starting with an honor code. It's simple and easy to remember, and it puts your values-in-use out there for your people to see.

An honor code for individuals is undoubtedly the first step to leader development whether the organization comes along or not. After all, individuals are the backbone of any organization, and most important decisions about the day-to-day operation of the organization are made by individuals at the middle and bottom of the group. And as Col. Banks said to me recently, "A good leader is a good manager by default. He or she is already doing things right, yes. And even though a leader can probably move the needle on individuals [toward values-congruent behavior], on large organizations? Maybe not."

Col. Banks continued in his interview with me, "A leader with no support at the top of the organization [a depleted environment] must get creative in order to foster values-congruent behavior. They will need to use persuasion and subtlety, tact and delicacy. You can't just let some fancy contractor allow his employees to ignore fall protection even when he brags about how he protects his employees. To get that one-to-one consistency between values and action, the leader needs both accountability and rule compliance and persuasion. And the leader can't just give up. It's a lot of work."

Another thing the depleted environment leader does to get the desired end state of values consistency is to view the situation through the "relevant cultural lens," in Bank's terms. This means that not everybody acts or thinks the same way. Banks relates a story about Iraq military officers always eating first, the complete opposite way that American officers act when they allow the enlisted men to eat first. But that didn't work in Iraq. Within a few weeks, discipline broke down among Iraqi soldiers, as the men dragged out dinner time, as was their tradition, longer and longer. The practice of the Iraqi enlisted men eating first was abandoned as a cultural mismatch. Nothing serious, but it resulted from not using the right cultural lens, or applying leadership techniques on this group just because they worked well on that group. They don't necessarily work.

In the steel industry, one of my graduates told me about how proud he was that he was working on behavior-based safety (BBS) plans and how he wanted the two unions at the plant to be intimately involved. Even though union and management sometimes squabbled about wages or time off, he was dead set that there wouldn't be any divisions on safety matters. Even in that challenging culture, his "cultural lens" had both sides working toward the same safety goals, and he made it work, even after 15 years. He planted "trust" as the most basic goal, and it worked.

Be, Know, Do

Part 1 of my Simplified Values-Based Leadership Model is for a lower-level leader who wants to begin to lay the fabric for leader development by starting with a simple accountability tool, the honor code. The leader tells subordinates exactly what is expected in those 13 words, with heavy emphasis on placing trust in each other (Figure 8.2).

In the second part of my simplified model, I have borrowed from Frances Hesselbein and Eric Shinsecki's (2004) book, *Be, Know, Do:*

FM 22-100

ARMY LEADERSHIP
BE, KNOW, DO
August 1999

Headquarters, Department of the Army

DISTRIBUTION RESTRICTION: Approved for public release; distribution is unlimited.

Figure 8.2 In a manual written by military and civilian experts, the Army teaches small group leadership in a simple model that begins with understanding our own motivations.

Leadership the Army Way. Do you remember me saying how impressed I was walking into West Point's Thayer Hall bookstore and seeing dozens and dozens of books about leadership?

Having waded through many of the available books on leadership theory, I am impressed with the basic honesty presented here in this book by recognized authorities from industry and the military side. It's really a nice blend.

You don't need a fancy title to practice the lessons in the model presented in *Be, Know, and Do*. You can be a shipping department director, the maintenance lead, the engineering project manager, or the safety director. But it is clear from the book that if you want to become an authentic leader, you'll need to step up beyond simpler administrative expectations. Even in the strictest line organization, it just isn't likely that upper management will object to inculcating your subordinates with high ideals. So let's look at the model presented by Shinsecki and Hesselbein and used widely among Army leadership trainers today (see Figure 8.3).

Be, Know, Do model of leadership

Be	Know	Do
Describes a person's innermost values	**Describes a person's competencies**	**Describes a person's decisions**
Values	Competencies	Decisions
Loyalty	Interpersonal skills	Influencing through communications
Duty	Conceptual skills	Influencing through decisions
Respect	Technical skills	Influencing through motivation skills
Selfless service	Tactical skills	Planning and preparation
Honor	Also: be competent in compliance issues	Assessing decisions ex post facto
Integrity	Also: be aware of industry best practice	Improving by developing subordinates
Personal courage	Also: be aware of trends and research in safety	Improving by building teams
Honor code: our department members do not lie, cheat nor steal, nor tolerate those who do	Also: be aware of trends in technologies and social media	Improve through personal growth

Figure 8.3 Even if upper management isn't interested whether you develop leaders in your own subordinates, this model is simple enough to be easily understood and easily practiced. (Adapted from Hesselbein, F. and Shinsecki, E. K., *Be, Know, Do: Leadership the Army Way: Adapted from the Official Army Leadership Manual.* San Francisco: Jossey-Bass, 2004.)

Be the ideal; personally represent the company's core values and standards; be the person with integrity and loyalty; exemplify the highest ideals and live the company's honor code; "walk the walk." And *be* not afraid to do what's right.

Know what you are supposed to do and when you are supposed to do it; be trained and seek regular competency and skill updates; be aware of trends in safety compliance and research; have the right technical skills to lead. Know what motivates your direct reports and how your own culture operates. Know where the power centers are and who are the gatekeepers to information.

Do the right thing each time and every time; have the interpersonal and communications skills to get your message across accurately and efficiently. Follow the rules and procedures even when you're by yourself. Remember the Spotlight Rangers who acted right only when their superiors were watching? They were always found out and ridiculed by peers.

I fully attribute the simple elegance of the Be, Know, Do model to Hesselbein and Shinsecki. I am merely borrowing it here as a reduced version of their original work and applying it in a different setting for safety leaders and engineer.

Storytelling, nonmaterial rewards, and personal courage

In a microenvironment, a charismatic leader has maximum impact using storytelling, as described in Chapter 7, and recognizing his or her people for doing the right thing. How should we reward workers who do the right thing? The answer is not intuitive but research based.

Frederick Herzberg is credited with the idea that motivators and nonmotivators at work are independent from each other and can be manipulated by the organization to reward or withdraw reward (Herzberg, Mausner, and Snyderman, 1959). Not intuitively, Herzberg's satisfiers at work were not things like money or a big promotion, but simple things like recognition for a job that contributed to company goals or taking on new responsibility. These are nonmaterial rewards for good work.

Dissatisfiers included doing too much paperwork and being forced to follow tedious rules, fairly obvious negatives in the workplace. But Herzberg found that material rewards such as a raise or a plum relocation acted as dissatisfiers, things that actually thwarted personal development. It turns out that nonmaterial rewards—things that cost nothing but mean everything—are much more motivating than material rewards.

Following Herzberg with my own research, I came to realize that nonmaterial rewards are almost always more powerful incentives than material rewards. This turns out to be particularly true among craft workers, who are often cynical or derisive about safety trinkets and especially money given for safe performance (there is probably nothing worse than a vice president handing out a safety trinket). In a 2004 study, we allowed scaffold operators to create their own checklist and perform their own inspections on a new and experimental scaffold at a large construction site in Baltimore. We allowed craft workers to set their own schedule and develop their own inspection parameters for the new scaffold as a reward for scrupulously following the lengthy and detailed instruction manual for the new and fairly complex scaffold system. The craft workers performed above expectation and the reward (work scheduling and unsupervised inspection) cost the company almost nothing at all.

This nonmaterial reward consisted only of allowing craft workers to set their own inspection parameters and do their work without heavy supervision. In turn, and according to the hypothesis that nonmaterial rewards would be motivating, the craft workers exceeded the number of inspections and the attention to detail (Winn, Seaman, and Baldwin, 2004). Notice how we conclude that even though workers said they wanted material rewards, they actually responded otherwise.

In our research, we found that after six months, the use of nonmaterial incentives significantly improved on-time delivery and completion rates of a special inspection form (both $p < .005$). In addition, a questionnaire with embedded critical questions showed that even though workers said that they preferred material incentives, we conclude that their behavior was changed by the treatment (incentives). We further conclude that the use of natural reinforcers seems to influence worker behaviors and perception of management's commitment to safety over the long run, even though workers still say that they prefer tangible rewards.

In theory, there are other nonmaterial rewards that can work to improve the frequency of values-congruency. This might include BBS principles applied to leadership. If we believe that behavior is a function of its consequences, then by rewarding values-congruent behaviors periodically with simple praise or public acknowledgement that the worker did the right thing when nobody was looking, we are increasing the likelihood of the behavior repeating and increasing the likelihood that others will emulate the original behavior. And that, my friend, can happen without upper management knowing or even caring. That, my friend, represents your leadership having a growth spurt.

Finally, I recommend using personal courage to demonstrate values-consistent behavior to subordinates. After all, that's what leaders do. But is *personal courage* something only for the movies and heroes? I think not.

Sometimes, ordinary but authentic leaders do extraordinary things to protect their subordinates because their actions are values-consistent. That means that leaders in even unsupportive environments can establish values congruency and incipient leader development at low levels. The board room need never know about personal courage at the department level.

There is a story of a West Virginian whose quiet personal courage should be a beacon to us all, and it was displayed in the most unsupportive, depleted environment you can think of. The story starts in Chester, West Virginia, known for a few things like the world's largest teapot. Chester is also home to Homer Laughlin China, a mile-long factory on the Ohio River that makes Fiestaware (ask your mother). Chester also has the last toll bridge that still costs a quarter (Figure 8.4).

Chester, West Virginia, is also the hometown of Mark McGeehan, an Air Force Academy graduate and pilot stationed in Washington State in the 1990s. He was serving as a copilot to a more senior Air Force officer, Bud Holland, who had been a celebrated fighter pilot before preparing to retire from the military. McGeehan and his crew worked under Holland on a B-52, the Air Force's huge, lumbering bomber that served then—and still does—as the country's pride and joy. So proud of the B-52 was the Air Force that they regularly held air shows to demonstrate to the public and to Congress how much a deterrent the B-52 could be when used in low-level maneuvers and at speed.

McGeehan saw right away that Holland was a risk-taker and discovered he had been written up, reprimanded, for taking chances with other crews (Figure 8.5). At one point, Holland said he wanted to be the first pilot to do a barrel roll in the giant bomber, something that had never even been attempted. Lots of other pilots laughed, but McGeehan knew he was serious.

Figure 8.4 Chester, West Virginia: home of the world's largest teapot and a story to match its size.

Figure 8.5 In a heartbreaking display of personal courage, Pat McGeehan refused to compromise the safety of his subordinates.

Still, Holland demanded respect because he was the most experienced B-52 pilot in the Air Force. A senior officer said, "Bud Holland has more hours flying a B-52 than you do sleeping." He may have been right.

One time, photographers wanted some special shots of a B-52 passing overhead with its bomb bay doors open, so they set up a series of cameras and equipment on a hillside so Holland could pass over at about 500 feet Just for effect, Holland blasted up the valley and barely missed the cameras and crew, passing less than 10 feet from the hillside.

Another time, Holland, now known in some circles for his rogue actions, exceeded the actual design limits of his B-52 in turning a tight circle over his daughter's softball game.

Col. McGeehan decided to go to his commander and ask that Holland be taken off flight status, a daring move that could have easily jeopardized his impeccable career record. But because this particular air show in 1994 was going to be visited by not only the highest level of Air Force officers including the Secretary of the Air Force and the local congressman for Seattle and Fairchild Air Force Base, McGeehan's request was refused. So Col. Mark McGeehan did something even more courageous and even more potentially damaging to his career. He took his own flight crew off flight status for any flights under Holland. If Holland wanted a copilot, McGeehan said only he would be going along; none of his crew would fly under Holland.

With dignitaries in attendance and McGeehan's own family and kids watching from their house on base, Holland made a few exceedingly low passes over the base. There are videos of these passes, and they are chilling to watch. But the most chilling video was taken as Holland's plane passed over the air field close the asphalt and circled around at an angle that again exceeded the recommended design limits of the aircraft.

That video shows the B-52 making a very tight left banking maneuver and way too close to the ground. The video shows Holland hitting the throttle hard to recover as he always did, but the eight seconds needed for the turbos to spool up weren't enough to keep the huge plane from cart wheeling into the ground in a huge fireball. The crash killed all aboard in full view of the dignitaries, including the local congressman, the Secretary of the Air Force, and McGeehan's own wife and three kids (Figure 8.6).

The photo of the crash used here shows a member of the crew ejecting right before impact, but the ejection was unsuccessful. There were no survivors.

Yes, there were investigations and inquiries, but these were all after the fact. Upper management knew full well that they had a rogue employee on board, but upper management did nothing to prevent him risking lives of his own subordinates. But one such subordinate did the most daring

AWRY: the B-52 exceeded authorized maneuvers and, after aborting a landing, lost altitude

Figure 8.6 At the Seattle air show in 1994, the lumbering B-52 could not emerge from a banking maneuver that significantly exceeded its design parameters.

things imaginable. The subordinate, Col. McGeehan, took it upon himself well in advance of the air show and, at risk to his own career, to call for the famed pilot, Holland, to be grounded. He also forbade his own direct-report airmen to fly with the cowboy pilot. He told his crew that if anybody had to fly with Holland, it would him, and only him.

That kind of courage inspires people, I hope, to reach higher, to do the right thing for their direct reports even under dire circumstances. But the story isn't quite over yet—if fact, McGeehan did inspire others, including his own preteen children.

After the crash, his family came back home. His boys played football at New Cumberland High School. His middle son, Brendan, went to WVU and graduated in physics and won a prestigious Goldwater Scholarship. His youngest son, Colin, recently graduated from the U.S. Naval Academy and is serving on active duty. McGeehan's oldest son, Pat, reached higher, too, graduating from the Air Force Academy like his dad, and later serving in Afghanistan and marrying a local girl. Later still, he started a small business in the West Virginia panhandle and finally running, and winning, office in the West Virginia House of Delegates. Their dad would be proud, I'm sure.

Figure 8.7 is one I have used for a while now in my class on practical leadership. I use it here because I just can't summarize the McGeehan story better than I already have in this slide.

We will modify to the Simplified Model Simplified Value-Based Leadership Model for use in depleted environments when we get to leader characteristics under crisis- and noncrisis-based situations in Chapter 10.

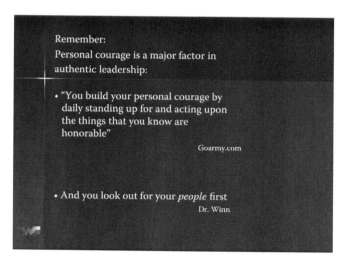

Figure 8.7 Personal courage and honorable actions are hallmarks of authentic leaders.

Later, in Chapter 14, we will discuss the debilitating effects of stress on people and the organization and add ways to keep organizational morale up even if you're being ignored by others in positions above you. We'll add the importance of acknowledging rituals. All of these are techniques a junior leader can use when nobody is looking, much less from the corner office on the 8th floor.

My central point so far is that upper management support for leader development is important, sure, but an enthusiastic leader way down in the system can create and foster a microenvironment—a mini-ecosystem—to start, support, and sustain leader development. I've seen it done, and I know you can do it to by following the simplified suggestions I have presented here.

It will cost nothing but your time and motivation to try.

Getting the depleted-environment model to work

How to get the "depleted model" established and working? Let's recall the theory from Chapter 7 in the section dealing with resistance to change. It was the James–Lange theory, remember? It flies in the face of conventional wisdom. Conventional wisdom says "attitude first, behavior later." But the James–Lange theory says otherwise. Let's revisit it for a minute.

According to that theory, if we were to encounter a bear while out for a walk in the woods, we run not because we are afraid. We end up with a physiological response first (a visceral fear of the bear), followed by an emotional response (being afraid), which are both caused by running on impulse. The running is reflexive and causes the other two responses, counter to intuition, which would suggest we are afraid first.

In this application, leaders mandate the required behavior first (wearing fall protection, for example), then they apply encouragement, peer support second, and finally, the attitudes will follow.

"Here is what I expect," says the leader, "and here are the penalties if the rules are not followed." Doing it this way and fostering attitude development is entirely transparent. There are no hidden agendas.

This isn't "do as I say or else." It's the leader showing how, telling why, and applying accountability. As the leader, *you demonstrate the behavior and hold subordinates accountable for doing it the right way.* The attitude will follow.

What does this have to do with leadership in the "depleted environment?" If leaders in an unsupportive ecosystem want the outcome every leader wants in a risky environment—values-congruent behavior—they start by modeling and requiring the behavior right away, add nonmaterial rewards, and work toward attitude change. The benefit lies in the fact that the values-congruent follower will model a much wider variety of safety behaviors once the attitude is established for the first behavior, and nonmaterial reinforcers are used. In our 2004 research mentioned earlier, strong and positive attitudes for fall protection led to strong and positive attitudes generally. Money didn't work even when employees thought it would.

To summarize, it is a sad probability that most of our current crop of safety managers and engineers will not work for one for the all-star leader development corporations. Most of you will work for organizations where the height of expectations is to be a good manager, the person who does things right.

The model I have offered allows a motivated individual to start acting like a leader, and providing leader development on his or her own.

Rather than despair that you might not ever work for the best corporations that do create systems to encourage leader growth, in this chapter, we face this reality but recognize that there is hope at the department level even when upper management is unsupportive. Using the best thinking available, I have proposed here that leader development can occur when the motivated individual follows these steps in this depleted environment.

Here's my Simplified Model for Leader Development in a depleted environment:

- Develop and share an *honor code* such as West Point's simple 13-word statement that "an employee in our department will not lie, cheat, or tolerate those who do." You can make up your own code, of course, but this becomes the floor for all other actions to develop leaders.
- *Be the ideal*; model the best (safe) behavior for coworkers.
- *Know your job*; advance your professional status; pay attention to details of safety and technical requirements of risk management. Know how to train others, or see that they get the best training possible.
- *Leader-developers do the right thing* each time and every time. They don't cut corners; they don't take risks for themselves or their subordinates.
- *Encourage storytelling among* workers who will pass along the lessons of near misses and close calls. Honest stories are always uplifting and motivating. Leader-developers will allow the stories to go where they will.
- *Create a system of nonmaterial rewards.* Acting in a values-congruent way should be recognized not with money or trinket rewards but by giving workers the right to change things. Attitudes will emerge after the structure and accountabilities are in place.
- *Demonstrate personal courage* even in adverse conditions. Modeling the right behavior even when it's hot outside or uncomfortable shouts "this guy knows what he is doing—this guy cares."

- *Make employees accountable* for safe work performance and measure their performance.
- *Require the safe behaviors first.* Remember the James–Lange theory; positive attitudes for many other behaviors will be created eventually.

Sure, management might never care whether you have an honor code in your department or a values-based system of action for developing leaders among your own subordinates. But even if management doesn't care or never even knows about it, isn't it the right thing to do?

chapter nine

Case studies in ethical considerations

As I write these chapters, I sharpen my focus on the importance of this material to my audience: young people who will be entrusted to watch out for the very essence of an organization, its people. Safety management students take a remarkable load of classroom work in math and science as it applies to identifying and abating hazards; these same young people will soon be responsible for plant security, fire protection, and a dazzling array of regulatory requirements. Engineers cram their four (and often five) years full of thermodynamics, heat transfer, and automated machine controls. Later, graduate civil engineers make sure the bridges and buildings are structurally sound. Mechanical engineers might design a sensor on an all-terrain vehicle that shuts off the engine if its carrying capacity is exceeded. Chemical engineers design piping solutions at a chemical plant to comply with process safety standards.

The future is indeed bright for safety professionals and engineers. I can't nearly place enough graduates into the open jobs I see come across my desk.

Indeed, the NIOSH said recently that the future is rosy for safety professionals: Two jobs await every graduate for the next 10 years (McAdams, 2011). NIOSH Director Dr. John Howard said in 2011 that the national demand for occupational safety and health services will significantly outstrip the number of men and women with the training, education, and experience necessary to provide such services. He went on: "Robust businesses are essential for U.S. economic recovery and growth, and in turn, safe and healthy workplaces are a vital ingredient of any successful business plan," said Howard. "The results of this NIOSH-commissioned survey suggest a troubling shortfall of professional expertise at a time when such services are most needed."

No wonder the *Wall Street Journal* and *US News and World Report* have shown engineering jobs taking the highest six or seven "in-demand" spots for decades.

Safety professionals and all engineering disciplines have academic curricula that are full of technical preparation, but there is precious little time, much less an actual course, devoted to the ethical environment in which designs and decisions and practices are made. I know this is true because I have taught now for 25 years in both camps. Deans and chairs say that they appreciate the need to discuss ethics, but their curricula are already full; there is no time or space for new material.

Sadly, I have heard it said by an engineering department chair in defending the crammed slate of courses his engineering undergraduates undergo that "this leadership and ethics stuff isn't engineering, anyway." He had no time for it. I might mention that the same chair is a great manager but will never himself be a leader with a vision for the future and that is the clear observation of his students and colleagues.

Don't accreditation requirements demand that safety professional and engineering students discuss ethics? Indeed, they do, but the standards just don't say how well, or how long, ethics is discussed. At a different school, another engineering professor, and chairman, agreed with me that instructing ethics is not given much priority at the highest level (known as "outcomes") by the accrediting agency (ABET) and not at all under the next level of importance (known as "objectives"). There are requirements, apparently, but the accrediting agency doesn't seem to notice that ethics receives very little attention out there in the real world.

Here's an interesting test to try: If you ask recent graduates, whether safety management students or engineering students, what is the first canon of their respective codes of ethics, I'll bet that none will know it. Ironically, the first canons of the ethical codes for the Board of Certified Safety Professionals (BCSP) and the NSPE are almost identical: Public safety and health are our highest priorities.

While these are my own observations, that ethics is not given much priority in university ABET standards (those governing engineering and safety management program curricula), others see it similarly. Our society as a whole seems in "moral meltdown," a phrase put so succinctly by Len Marrella in his 2009, third edition, book, *In Search of Ethics*. Marrella anecdotally points out some common and accepted but clearly unethical practices in today's United States, which might include the following:

- Cheating in college is considered OK.
- Politicians flip-flopping on important issues is OK.
- Pop culture icons are routinely arrested and that seems OK.
- Parents using marijuana around their kids is OK.
- Taking these leftover hard drives from a company is OK.
- Nobody knows about the blood-doping, so my medal is safe; it's OK.
- As long as nobody got hurt and nobody got caught, it's OK.

The short and easy answer is that these behaviors that Marrella mentions are patently not OK, but worse, we all know they're not OK. Where, then, is the root of the problem with growing acceptance of unethical behavior?

Remember a chair in my college of engineering said, "It's not engineering anyway so it isn't important."

Are ethical considerations intertwined into safety or engineering and do these considerations also interact with leadership? The short answer is "yes," but let's look into these questions using examples.

The safety professional acts in an unethical way when he or she overlooks an unsafe act by a worker who is oiling a moving drive chain without locking out the engine. That act is easily judged to be against safety principles and passing it by should be against the safety professional's own ethical principles. It's simply not OK to skip the lock out procedures, and it isn't OK to overlook the subordinate who does.

But what about applying for travel reimbursements that he or she didn't earn? Or purchasing a large and expensive stamping die because her brother-in-law is the vendor? These actions didn't hurt people and the manager isn't likely to get caught, so they're OK, right? Somehow, these actions don't quite cross the line as unethical but they actually do cross the line because ethics as a whole is disregarded precisely by the very people who should teach it.

Engineers are no less challenged in their daily work to do the right thing. They act in an unethical way when they sign off on a design drawing done by their intern. They act in an unethical way when they talk authoritatively about a potentially hazardous chemical reaction even though it is out of their range of expertise. These didn't hurt people, and they aren't likely to get caught, so they're OK, right?

Just like Collins said earlier about the military: In the business of safety and engineering, there are no do-overs, and people get hurt when their leaders get it wrong. Regardless of the ABET standards, which, to me at least, seem to minimize the importance of ethics to our undergraduate engineers and safety students, I think students need a much fuller and richer presentation of this important topic.

Ethics, morals, and values: How are they different?

Let's define our terms first using Marrella (2009) as a guide. *Ethics* is a system of moral principles dealing with human values and moral conduct. It implies "duty" and "the ideal human character." It is essentially values and morality applied to groups and decisions that are made in a group context. Recall that in our discussion of characterizing a profession, I said in Chapter 3, that codes of ethical behavior are issued among members, suggesting appropriate behavior as judged among peers in the profession.

Morals are acts, or conduct, that are made on the perception of right and wrong judged by a person's conscience. A person has a natural sense of moral principles. Here is a key point: Morals are owned by the person, while ethics are principles judged by a group.

Values, as we defined in Chapter 5, are constructs, high ideals, a principle, or standard that is worthy for its own sake and needs no explanation. Values, like morals, are owned by the person.

Integrity is a behavior that is based on adherence to a code of conduct, being honest with one's self about his or her own behavior.

Character is having the courage or stamina to act in accordance with a person's set of individual values.

At this juncture, we need a treatment of ethics because in the full development of a values-based organization and values-based leadership, an ethical guide is a public declaration of what our larger group thinks is important. Some codes, or canons as they are called formally, are long and some are brief. But notice in the examples I will use here soon, some real similarities exist across canons of ethics, not the least of which is the almost perfect congruency of the BCSP and NSPE first canon about the importance of the public health and safety.

Other similarities across canons involve the crucial matter of trust. Engineering clients, for example, will almost always know less about the subject, let's say, of the hydro-geology of a proposed building site, than the engineer does. The client trusts the engineer's judgment, and in return, the engineer has a huge responsibility to meet the expectation of trust for the client or the public at large. The township supervisors may say the site's hydro-geology is suitable, but the client and the public trust the professional engineer more.

The same exists with safety professionals: Craft workers will almost always know less about a subject, let's say, the need for a local exhaust when welding inside a truck body, even if the welder says, "I don't need it—I'll just be welding for a minute." The company CEO trusts the safety professional to ensure that the welder gets a fabricated exhaust system up and working even if the welding duration is brief. Trust doesn't come out of a text book. It is instructed and carefully built upon the foundations of ethics, mortality, and values.

When it comes to trust, I used to tell my young engineers precisely the same thing I told my safety students about trust.

You have a high standard of trust to live up to. First, people who don't even know you will tend to trust your judgment once they know you are a professional in the respective field. Second, you will have to protect that trust by acting in accordance with group norms. And third, acting aberrantly with regard to the trust may immediately hurt a person or persons involved, but there is also hurt experienced by the profession itself when this trust is abused.

Even if we never got much further in a discussion of ethical considerations shared by safety professionals and engineers, two major ethical considerations are shared by the respective communities. First, both professionals have a moral obligation that comes before any compliance or regulatory obligation not to be negligent and to use reasonable care in making decisions. Second, both groups of professionals must warn about hazards they are aware of.

Elizabeth Stephan is a trained chemical engineer teaching at Clemson University in South Carolina. She and her collaboration team have written *Thinking Like an Engineer: An Active Learning Approach* (2011), which is used in introductory engineering classes at many top schools around the country, including WVU. This book was brought to my attention by one of WVU's freshman engineering instructors, Dr. Ordel Brown, someone I have high regard for as a teacher of young people and engineers in particular.

Stephan offers a simple algorithm for testing whether an intended action is ethical or not. It's easy to practice and easy to remember. Using her method adapted from *Thinking Like and Engineer,* we test the action using four perspectives, and remember, these are seat-of-the pants tests, nothing scientific or research based. There is no need for it, as we'll see.

Is the intended action ethical from the perspective of its consequences? Might there be losses of trust, a conflict of interest, or even a loss of potential business if the action is performed? The greater the negative consequences, intended or otherwise, the more likelihood of it being unethical.

Is the intended action ethical so far as its intent? That is, should everyone commonly perform this action in other situations as I intend to do? Would I like to be on the receiving end of this action? Could this sort of intent or purpose be built into law? If the answers to these questions are "no," it is pretty easy to see that the intended action is unethical.

Is the intended action congruent with the behavior of a person with a respected character? That is, what would a person of known and established good character do in this same situation? Moreover, what would the consequences of the intended action do so far as changing the actor's character? If it changes the reputation in a negative way, the intended action is probably unethical.

Would the intended action elicit the same responses from other people? Would other people you know and trust have fairly uniform responses to the intent? If the answer is uniformly negative, then the intended action is probably unethical. Even if the response set is mixed, the actor might still be advised to avoid the action just because of the appearance of potential problems.

Once the intent of your proposed action is carefully examined with respect to the foregoing questions, or a similar set that you develop, then it's much safer to act in confidence that your decisions are ethical by the standards of your group, or else you should avoid the planned course of action.

Usually, the individual canons of an ethics code are fairly well defined until we throw case studies at them. Things get a bit murky when we put them into terms of the real world. In fact, let's apply the foregoing tests with seven examples I use in class. These are examples that I developed for my engineering classes, but notice that they can be used immediately by safety professionals, too (Figures 9.1 through 9.7).

Ethics case study example 1

> Dilemma: Engineer Fred is a building inspector for a large city. Fred's brother, Ned, is a builder, but his apartments have a reputation of having some serious wiring-safety problems.
>
> Discussion: Because they are brothers, there is a conflict of interest that Fred must disclose. He must not inspect Ned's building just so there is no appearance of favoritism.

> Source: "Engineers shall disclose all known or potential *conflicts of interest* to their employers or clients by promptly informing them of any business association, interest, or other circumstances which could influence or appear to influence their judgment or the quality of their services."
>
> Copyright 1997 National Society of Professional Engineers, Fundamental Canon 4.A.

Figure 9.1 Conflict of interest.

Ethics case study example 2

> Dilemma: Petroleum engineer Ernie inspects a piece of land for his sister and her husband, who want to buy the land to build a house. Ernie sends a report of his finding of natural gas deposits under the land to the couple and sends a copy to the real estate agent.
>
> Discussion: Ernie has breached confidentiality of the couple and in turn reduced their ability to buy the land at a good price.

> Source: "Engineers shall not reveal facts, data, or information obtained in a professional capacity without the prior consent of the client or employer except as authorized or required by law."
>
> Copyright 1997 National Society of Professional Engineers, Rule of Practice 1.C.

Figure 9.2 Client-consultant privilege.

1. Hold paramount the safety, health, and welfare of the public.
2. Perform services only in areas of their competence.
3. Issue public statements only in an objective and truthful manner.
4. Act for each employer or client as faithful agents or trustees.
5. Avoid deceptive acts.
6. Conduct themselves honorably, responsibly, ethically, and lawfully so as to enhance the honor, reputation, and usefulness of the profession.

II. Rules of Practice
1. Engineers shall hold paramount the safety, health, and welfare of the public.
2. Engineers shall perform services only in the areas of their competence.
3. Engineers shall issue public statements only in an objective and truthful manner.
4. Engineers shall act for each employer or client as faithful agents or trustees.
5. Engineers shall avoid deceptive acts.

A young professional's generalized code of conduct: A set of ethical canons you can use at the department level

Are you not a member of a recognized profession yet? Are you thinking about entering a profession where there is no established code of ethics that you can find? No worries, feel free to use the following set as long as you (ethically) attribute it. For purposes of this book, I have borrowed some material myself, and I attribute it below.

Here is a *generalized code of conduct* for both safety professionals and engineers that you can adopt in your first management job. Remember that a code of ethics is based on how the profession sees itself and wants to be seen by the public at large. And remember also that a code of ethics is an expression of group values, not necessarily individual values.

Members of my engineering or safety and health group will:

1. Protect the public safety, health, and welfare above all else.
2. Perform duties only in your area of expertise.
3. Be truthful and objective.

4. Behave in an honorable and dignified manner befitting professionals.
5. Continue to learn new technical skills.
6. Provide honest and hard work to any employer or client.
7. Inform authorities about any harmful or illegal or hazardous activities you become aware of.
8. Become involved in civic and community activities; volunteer your expertise as a service to your community.
9. Protect the environment.
10. Do not accept bribes or gifts that would interfere with your judgment as a professional engineer or safety professional.
11. Protect confidential information, plans, patents, or designs of your client or employer.
12. Avoid obvious conflicts of interest and take pains to avoid even potential conflicts.

Source: Winn after Holtzapple and Reece (2006)

To summarize about ethics, individual morals come before ethical consideration shared by professional groups. Whether in safety or in engineering, professional practitioners have created sets (canons) of statements to guide individual behavior in the respective group. Such is the deliberation given by the top groups in each field that public safety and health are given the very highest priority.

Add this generalized code of conduct to your own honor code, and you have the basis of a solid culture of trust and integrity, even if upper management isn't looking or doesn't care. These are your subordinates we are talking about: How you take care of them and what your expectations are determine the culture you construct in your department or unit. It supports values-in-use, which in turn supports authentic leadership.

Plagiarism and consequences for professionals

Before leaving the discussion of ethics, we should take a minute to discuss plagiarism, its consequences for young safety professionals and engineers, and how to avoid them. As all of us pretty well know, plagiarism is to "steal and pass off the ideas of another as one's own, or to use another's product without crediting the source" (Merriam-Webster.com). We've been around this concept since college and probably since high school. It isn't new. But please don't just skip this section because you think you know all about it. There are some new twists that young tech-savvy professionals need to be aware of.

You might have considered plagiarism to be something you talked about in high school with not much applicability for the future. I am afraid that is not only incorrect, but also the consequences—and frequency—of plagiarism are actually bigger than ever.

Among what's new for people entering the workforce, especially since I've been in college, is the idea that plagiarizing is entirely too easy to do today. Given the ease with which one simply can Google a term and come back with pages and pages of material, it's way too easy to cut and paste somebody else's ideas into your own document. But even Pleistocene-aged professors can be pretty clever, too. It's become pretty easy to check for plagiarism by simply Googling passages to see where else they come up.

So what's really new for entering professionals even if the foregoing is not new? Is it "moral meltdown" as Morrella described previously? Is it moral relativism? Do people simply not care? In brief, no is the answer to all of these.

Dawson and Overfield's (2006) research shows that students are simply unclear about the distinctions among collusion, plagiarism, and permissible group work and that science students, in particular, were not aware of the boundaries between permitted work and otherwise. There is nothing sinister or conspiratorial about young people and plagiarism; it's just that they don't know what the boundaries actually are.

Looking for more research on plagiarism, I spoke with Marian Armour-Gemmen and Mary Strife, who are both exceedingly helpful professional librarians at WVU. Along with our associate dean for freshmen engineering, Dr. Robin Hensel, they presented a peer-reviewed paper to the ASEE, which not only addressed technical writing needs for engineering freshmen but also summarized the research literature and issues associated with plagiarism and collusion among technically-savvy students. I would direct the interested reader to their ASEE paper for a full and excellent literature review (Strife, Armour-Gemmen, and Hensel, 2012). These authors concluded that, with proper instruction, freshman engineering students did better understand plagiarism and how to avoid it, know how to evaluate an article, were able to appropriately cite an article, were familiar with four source databases for engineering research, and were able to identify the four types of intellectual property. In-class exercises, readings, and quizzes were geared toward these outcomes. So, education about plagiarism seems to work.

But there are some serious repercussions for not doing things right when citing another author's work fairly. I found an interesting 2012 article by Erin Schreiner, a freelance writer for the online series titled "eHow." (See how easy it is to use attribution? I just did it.)

Schreiner says that the least problematic consequence of plagiarism in the corporate world is *employer sanction*, which might include a verbal or written reprimand for minor misattribution, ranging up to a formal warning with more serious consequences for a second instance. Paper reprimands don't fade away quickly. And they can be passed through internal company channels widely and quickly by purring them in electronic form.

At the second level of reprimand, she says, the *employee is terminated* (fired) for misrepresenting someone else's work for his or her own, and maybe involving the employer in a lawsuit over it. Unlike in college, where a plagiarized paper would earn an F grade, the employee here is "on the street," with no unemployment benefits due to the termination action, and once again, spreading the word company-wide easy and fast.

Now, Schreiner says we have some really far-reaching consequences for the new employees, and that is *blacklisting*. It isn't clear in the brief article if this is common practice or whether it is even legal, but what if it is? Surely, in tight-knit industries, blacklisting is done. It means that the employee, highly trained in the technical and engineering aspects of the company, is no longer able to work in the very industry for which he or she has trained.

While that may seem like as bad as it can get, Schreiner says that under some conditions, the plagiarizer can be sanctioned financially in court with *civil penalties*. The penalty, if any, depends on the amount and the type of material and probably also the intent (although she does not speak to intent).

Without a doubt, working in industry isn't like college where you can leave your plagiarizing F on your transcript and move on. In industry, there may be no place to go. What can a young professional do to avoid it?

What's wrong with just Googling it?

Using somebody else's work without attributing it is unethical, and it can have far-reaching consequences for your career. Why not simply provide internal citations just as I done throughout this text (Schein, 1992, for example). It's always better to anticipate problems with the appearance of plagiarism by using more citations that you think you need. Cite the source even if it's our old friend Wikipedia.

In addition, you can use bibliography-type citations at the end of the paper (almost nobody uses footnotes at the bottom of the page anymore). You can use styles including Modern Language Association (MLA), American Psychological Association (APA), and even the Chicago/Turabian style, which is still used in some industries. These aren't difficult to use, and even trying to use them is a good faith effort at avoiding even the appearance of plagiarism, which may be more important anyway.

Bottom line: Be aware of the long-lasting consequences for plagiarizing or even the appearance of it; use internal citations whenever possible; and for longer papers of publications outside your industry, routinely cite all sources at the end of paperwork. A stumbling, good faith effort at full disclosure is better than even the appearance of plagiarism.

Figure 9.10 Air Force officer Jeremy Slagley, trained as an environmental engineer, is now a professor at Indiana University of Pennsylvania teaching safety and health.

Are morals relative? A dialogue for today's professionals

Jeremy Slagley, PhD, CIH, CSP, is shown in Figure 9.10 when he was still a commissioned officer in the Air Force. He is now an assistant professor in the Department of Safety Sciences of Indiana University of Pennsylvania. He retired from 20 years of active duty as an Air Force Bioenvironmental Engineer officer engaged in industrial hygiene, environmental health, and emergency response. His final assignment was leading a Flight of 44 military and civilian EHS professionals to protect the health of a depot base population of over 25,000 people. Dr. Slagley, an engineer and a safety professional, writes the following essay on why morality, and hence safety, is not relative.

> *Moral relativism* is the viewpoint that there are no universal moral truths. In effect, what one person holds as the right thing to do is not necessarily the same standard for the next person. What right does one person have to hold another's actions up to a moral standard that the other individual does not share?
>
> While this "live and let live" notion is very popular, it falls apart rapidly when someone attempts to defy gravity (a universal standard), or mathematics (I think two plus two should be seven). A typical

argument for moral relativism is that it is not someone else's business to judge one's behavior.

However, it is certainly everyone's business when your behavior can directly and significantly affect safety and health. Moral relativism cannot be accepted practice in the industrial workplace where risks abound—US military chaplain William T. Cummings said during the Battle of Bataan in 1942, and rightly so, "there are no atheists in foxholes." Workers, supervisors, business leaders, and safety professionals must adhere to a *common moral standard*—ethics. Why? Because protecting people, property and business efficacy can't vary by supervisor or day and worker. How could it be moral to protect only certain classes of people? It couldn't. *That's why moral relativism has no place in safety or engineering,* but let's explore this a bit more.

There is quite a bit of activity in the field of ethics in recent years. Every field of human endeavor teaches ethics, demands ethics, enforces ethics. We have also seen in the preceding chapter that every professional organization, engineering or safety, has its own code of ethics. Yet it is evident from the evening news, Fox or CNN that individuals in all fields of endeavor, at all levels from the entry level employee or new recruit to the chief executive, government leader, or general officer, have failed to "do the right thing." The Challenger space shuttle's engineers knew it was wrong to use a certain o-ring when the temperature dropped below freezing. Ford Motor Company's safety experts knew that placing the Pinto's gas tank at the rear of the car did save space, but it was a fire hazard. Merely having a code of ethics does not guarantee compliance.

While the rule of the day in the western world is "what's good for you is good for you" and "do whatever feels good" the only apparent limit or check on action is whether that action hurts someone else who has not consented to it. That's the "relativism" part in full flower. However, even hurting oneself impacts others depending on that individual. Moral relativism—having safety or engineering ethical principles on a sliding scale—is insufficient in any area where health or life is concerned. Joseph

Cardinal Ratzinger sagely warned the College of Cardinals of the Catholic Church in 2005 as they were entering the conclave that would elect him Pope Benedict XVI, "A dictatorship of relativism is being formed, one that recognizes nothing as definitive and that has as its measure only the self and its desires." (Catholic Exchange, 2005) Can an organization simply exhort their members to, "do the right thing" and hope it gets done? Not likely.

Manners vs. dogma

"Do the right thing" works as long as everyone agrees on what the right thing is. This is a simple definition of morals (Kidder, 1996). Lord Moulton gave an even better exposition of morals as, "obedience to the unenforceable" (Moulton, 1924). He described a continuum of the interplay of rules and behavior. At one end is positive human law (let's just say: on the right side for simplicity) which one should obey or there will be consequences (if caught). At the far end (let's say the left side) is free choice, where one has complete liberty to follow his whims regardless of any consequence. The libertarianism of absolute free choice, our left side, is one of the fruits of moral relativism. The realm in the middle Moulton describes as "manners," where one self-regulates because to act otherwise wouldn't be proper. But safety professionals and engineers know the importance that safety hinges on *workers self-regulating when no one is watching*. This culture would go beyond co-workers watching out for each other and really approaches the goal of a safety culture. But self-regulating presupposes the understood standard for proper behavior in that situation. It becomes apparent that moral relativism, which rejects any universal standard of proper behavior, is completely inadequate for workplace safety. The far right side of our continuum, and also the far left, just can't work in safety.

For a safety professional or project engineer who cannot be there to observe every critical behavior across the workplace, it is essential that every employee from the entry level to the supervisor "do the right thing" every time especially when no one is watching. The current standard of safety practice consists of job safety analyses, critical behavior inventories, fault tree analyses, and similar assessments of possible sentinel events. While these systems work well in small applications, they quickly mushroom in any complex work area or system—much less internationally where cultures and systems are greatly different—so that few of the authoring engineers or safety professionals completely conceive of the entire assessment and consequent guidelines. Such effort becomes very slow to adapt to change, which is often the most hazardous time of a process.

Furthermore, the actual worker may be overwhelmed by the analysis and either not be aware of the guidelines, or simply not bother with safety controls when they are inconvenient. The safest solution involves workers who have manners. That is to say, safety demands workers who consider safety as they encounter new and evolving situations, and of course, this is not moral relativism in any way. If it is a situation where controls are prescribed, they adhere to the controls (even when no one else is there). If it is a new situation but a familiar hazard, they take appropriate control measures and ensure others do likewise. If it is something entirely new, they stop the process and contact a safety professional or the project engineer to assess and advise. There is nothing relative about that.

Can safety be like "polished shoes?" Let's hope so

It is not in man's nature to accept "big brother" safety professionals watching over them at all times. That's the notion of "safety pro being safety cop," and we have moved away from that simplistic notion long ago.

But we know that man is created with intellect and will. Workers can know what appropriate behavior is in a situation and use their will to do that behavior and avoid inappropriate behavior. By implementing a common set of moral standards (an honor code), workers at all levels can exercise their intellect and will and eventually internalize those standards. Then they will be able to better use their manners and act appropriately when no one is watching or in new situations. This is a key step to move from the untenable goal of safety professional or engineer assessing every possible hazard and staying on top of it at all times toward a safety culture where the workers, supervisors, and corporate leaders all watch out for each other and demand "manners" from each other. No safety cop. No relativism.

West Point was the only place where I could drop a dollar bill on the ground before a class and look for it hours later and find it in the same spot after hundreds of highly polished shoes had tramped right by. That kind of adherence to a common moral code was essential for training new officers to lead our military against any enemy and be entrusted with the power of life and death. Industrial workplaces have similar high-risk situations and similarly could benefit from a common moral code. Dr. Winn has said that, and I highly concur. And we also agree that having an honor code is not a thing of the past: it's actually a thing for the future of safety if we let it be. Even in huge bureaucracies, an honor code can thrive at the department level. Simple manners can thrive in programs or units way beneath the surface if we have caring safety professionals and engineers, and the enthusiastic support of leadership at all levels. The cost is low, and the impact is unbelievably big.

One last word on moral codes is that they apply at all times in all spheres. We should apply the moral code to all aspects of our lives and demand the same of our coworkers at all levels. If the CEO is not maintaining high moral standards in his private life, how can he be trusted to lead the company? Consider General David Petraeus, arguably one of the greatest modern military minds. As director of the Central Intelligence Agency, it came to light that he had an extramarital affair with his biographer. He resigned in 2012 in disgrace. When it comes to safety and health, an individual who breaks their solemn oath to their spouse calls into question their trustworthiness to protect workers' lives on the battlefield or even the workplace. Similarly, if a worker has to trust his friend to de-energize and lockout a process for maintenance, but that friend is untrustworthy in his private life, the worker is hesitant to do the task.

The "live and let live" code of moral relativism is perfectly fine for food preferences and favorite colors. For safety and health it cannot be, "what's good for me may not be what's good for you." It must be, "what's good is good, and what's evil is not good." In this way, morality and hence safety cannot be relative to the situation or the condition or even the person.

References

Catholic Exchange, news. (2005, 19 April). *Ratzinger Warns Cardinals Against 'Dictatorship of Moral Relativism.* Retrieved 2005, from http://catho licexchange.com/ratzinger-warns-cardinals -against-dictatorship-of-moral-relativism/ February 23, 2013.

Moulton L. (1924, July). "Obedience to the unenforceable: Law and manners," *The Atlantic Monthly.*

Kidder, R. M. (1996). *How Good People Make Tough Choices: Resolving the Dilemmas of Ethical Living.* New York: Simon & Schuster, Inc.

In this chapter, we have examined how ethics, morals, values, and integrity are defined and used among career professionals. Ethical codes of conduct for many professional communities of engineers or for safety all have a common thread about protecting public safety and health first and foremost. We examined how plagiarism can unfavorably impact a young professional's career despite the ease of borrowing material without attribution. Finally, we examined whether there can be such a thing as "moral relativism" and concluded that in the fields of safety or engineering, there can be no rules that are good for some and not for others. In the same way that two plus two does not equal seven, there are certain, immutable truths about caring for others in ways and language that don't change.

chapter ten

Crisis and noncrisis leadership models

The following material represents the real meat of the book. It is an exploration of a wide range of leadership theories using the most current literature available. The coverage is not exhaustive, but only a treatment broad enough to suggest that *there is a lot of work available on leadership* to explore. But first, a word about using military examples.

Why should we study how the military teaches leadership?

It has been my position for purposes of this book that safety leaders or engineers have much in common with military leaders, despite some real pushback from the outside (including some prominent association officials) suggesting that I should not use military examples ("too different," "the military has a captive audience," and so forth).

But my position is this: certain kinds of civilian and most military leaders share strikingly similar objectives: to preserve and protect the people, property, and efficacy of their respective organizations. Let me explain further and offer some support for my ideas about how we can learn about leadership and organizational behavior from out military brethren before we get started on a discussion of prominent leadership models.

Reason no. 1: The experts say we should pay attention

As Jim Collins (2006) points out in *Leadership Lessons From West Point* (Figure 10.1) we are both engaged in occupational pursuits where making a bad decision means that *people can die.*

A military leader, just like the safety or engineering leader, has a clearly defined objective, and in accomplishing the objective, he or she will have to protect the people and property assigned to the task. He or she will have to accomplish the military objective within ethical and regulatory bounds, taking care to see that the soldiers are well equipped and, above all, trained in the particular skill set needed. The military and also the civilian leader studies particular hazards associated with particular mission accomplishments and prescribes engineering methods first, then

Figure 10.1 I have used this textbook in my graduate-level class on professional development. It continues to be one of the best available textbooks on leadership and leader development.

administrative methods, and finally, personal protective equipment to abate the hazards. And if we cross out "military" and insert "safety professional" or "project engineer" and replace "soldier" with "craft worker," we see precisely the same process is done by a platoon leader in charge of 30 soldiers, or a project engineer responsible for 30 people building a courthouse. Same mission, same methods.

I am surprised as to why the safety and engineering professions have ignored these linkages. Military psychologists have been at the forefront of areas of organizational behavior, as I will soon show in this book, yet their research has been rarely cited or applied outside their world until very recently. For example, see the very interesting article about what the country's best known author on leadership, Jim Collins, recently discovered in "The Re-Education of Jim Collins" in *Inc.com Magazine*. In just one example, Collins points out how leaders in the military acknowledge failure in both academics and on the playing field, sometimes every day. As Collins points out, the cadets and faculty alike colloquially call it "embracing the suck," but Collins goes on to point out how this simple maxim is essentially missing from industry and is not cited even in his own influential books because he has only recently come to these conclusions—after his books became popular a decade ago. In a 2013 article, author Bo Burlingham, editor-at-large of *Inc. Magazine*, says of Collins, arguably the most widely read author on leadership theory in the past two decades:

> Glancing around the [West Point] gym, Collins could see numerous other cadets struggling with various obstacles; some of them were not much farther along than he was. Most of them had at least one or two other cadets standing nearby, coaching, critiquing, and cheering on their compatriots.
>
> That struck Collins as interesting. West Point is a highly competitive place. Every cadet wants to do the [physical] faster than his or her peers. Every cadet also is extremely busy. Yet these cadets were taking time away from their studies and other duties to help their friends get through the course.
>
> But Collins, of course, is best known for pondering the secrets of organizational, not personal, success. So what do these West Point revelations mean for company leaders whose shelves are lined with Collins's books?
>
> He sees a number of useful lessons. First, "If you want to build a culture of engaged leaders and a great place to work," he says, "you need to spend time thinking about three things."
>
> - Service to "a cause or purpose we are passionately dedicated to and are willing to suffer and sacrifice for."
> - Challenge and growth, or, "What huge and audacious challenges should we give people that will push them hard and make them grow?"
> - Communal success, or, "What can we do to reinforce the idea that we succeed only by helping each other?"
>
> Collins says he has observed these principles in action in a number of companies he has studied, at least during their best years; including IBM, Apple, Johnson & Johnson, Southwest Airlines, and Federal Express.

Collins also points out in this same article how he discovered that "selfless service" is something to which cadets and faculty are all dedicated. Service, he says, is a "cause or purpose we are passionately dedicated to and are willing to suffer and sacrifice for." He says he discovered that industry could also benefit from having employees who do service projects on Saturdays once in a while to link them to the community, and especially if civilian leaders provided "audacious challenges in order to push people harder and make them grow, and communal success opportunities." While a few companies he studied did this already, such as IBM, Apple, and Johnson & Johnson, most did not, and more compelling, almost no companies do this at anything but the senior levels.

"Great leadership at the top," says Collins, "doesn't amount to much if you don't have exceptional leadership at the unit level. That's where great things get done," as he distinguishes what the military does now and what industry could learn from.

About the same time I discovered Collins' article I mentioned previously, I had just finished drafting and sending off a manuscript I had written with West Point's Col. Bernie Banks for ASSE's Professional Safety. It is titled "Recognizing the Importance of Military Organizational Research in Developing Future Safety and Engineering Leaders" (*Professional Safety*, January 2014). Our manuscript makes similar points as Collins makes: What are the useful techniques the military have studied and now use that are missing in industry? I was told to expect "pushback from industry," but on the contrary, there has been only a little.

One thing Banks and I point out in the ASSE article that the civilian industry could benefit from is *experiential training* (see Chapter 12). Some call it hands-on training, but either way, it amounts to intense practice in real-life situations and then practicing it again, then again with opper management as active participants. Imagine: Clearing a weapon jam during a firefight is just about as important when the chips are down as clearing a jam in a fire nozzle during a full-on mine fire. We better get it right the first time, but sadly, most civilian training is only mildly experiential and learned on PowerPoint slides and classroom settings. That could change if the civilian world pays more attention to what others are doing. Jim Collins has paid attention, for one.

So don't just take my word for it that we might benefit from studying military leadership models. If you ask who does the best job in developing leaders, ask Jim Collins, the author with the most influential contemporary books on leadership on the market today, or Peter Drucker, one of the country's foremost theorists on management theory and practice, or even Jack Welch, the former CEO of General Electric and often called the best leader developer the United States ever had (Figure 10.2).

Leaders will need to act differently when they are in a crisis situation: Lives may be at stake. Managers, again by definition, will never be prepared in this way. There is often a single moment when a manager decides "enough is enough" and takes matters into his or her own hands, putting lives and safety first and their own safety—or career—at risk.

Let's examine two cases where an "average Joe" reaches the point where he or she makes a conscious decision to protect subordinates even in unconventional ways. The decision and the acting are hallmarks of true leadership at work in the crisis mode.

Case 1: GM ignition switch

A leader may even put his or her own job in jeopardy by calling attention to the truth. For example, consider the engineer who blew the whistle on the General Motors (GM) car ignition lock failures, something GM apparently tried to hide for years until some engineer had the gumption to stand up and tell the truth.

On June 18, 2014, Bloomberg.com tells exactly this story (http://www .bloomberg.com/bw/articles/2014-06-18/gm-recalls-whistle-blower-was -ignored-mary-barra-faces-congress).

> A General Motors employee who continued to notice safety issues in vehicles and spoke up was transferred to a job where he was instructed to "not find every problem that GM might have," according to a Bloomberg story posted today.
>
> Courtland Kelley, now 52, has worked at GM for more than 30 years, originally as a safety inspector, then in 2002 as brand manager for the Chevrolet Cavalier—the Cobalt's predecessor—and the Pontiac Sunfire.
>
> He declined to be interviewed for the *Bloomberg Businessweek* cover story. He remains employed by GM in its quality organization and can be considered part of the quality team, a GM spokesman said.
>
> Though Kelley brought legal action against GM, he remained employed by the company, being moved from one position to another. It is not immediately clear what Kelley's current title is.
>
> "He still has a job—he doesn't have a career," Bill McAleer, a former GM employee, told Bloomberg.

Choosing to do the right thing under extreme conditions can cost promotions and advancement, but committed individuals would do the same thing again because of spinoff improvements and protecting lives. At a minimum, GM CEO Mary Barra vowed that she would "fix GM's passive culture and encourage employees to speak up," according to the Bloomberg article.

Case 2: An officer protects his men

The same kind of crisis-mode will cause true leaders to choose a difficult path. *The New York Times* (http://www.nytimes.com/2004/05/27/world /struggle-for-iraq-interrogations-colonel-risked-his-career-menacing -detainee.html) ran a moving story in 2004 about an Army colonel who was tired of seeing his own men ambushed night after night in Iraq; he knew that his patrol plans were being leaked but he didn't know by whom. It turned out that a turncoat Iraqi policeman was giving the plans to local insurgents, who used the information to attack U.S. soldiers. Col. Allen B. West decided to use crisis-mode interrogation tactics to force the traitor-policeman to talk and fired his gun next to the man's head, causing no injuries but effecting a complete confession. The action stopped future attacks, and even though West's actions caused some outcries of abuse, not everyone saw it that way.

Colonel West was ushered out of the military for not following interrogation protocol. The former colonel is a high school teacher now and has no regrets about his decision to protect his men under those crisis conditions. In looking out for his men, he made a decision. "The fact is, I made a choice, the choice had consequences and I accept that," says the article.

What is a Level 1 crisis and can it be survived?

Klann has done an exemplary job of making the notion of *crisis* understandable by arranging them in a hierarchy, or levels. Soon, we are going to use this hierarchy at some length when we discuss models of leadership and how leaders are called upon for different attributes when a crisis emerges. The following are attributed to Klann in the same 2003 book I mentioned previously, and I have adjusted them to suit our needs in safety management and engineering project management.

In a Level 1 crisis, according to Klann, the organization undergoes "public embarrassment and mission success is threatened." Outcomes might include being written up in the newspaper in an unflattering way. For example, a headline reading "EPA Traces Chemical Release in River to Major Local Manufacturer" would cause a temporary distraction of goodwill and even company mission, but not on a permanent basis, unless, extending the example, the company is discovered releasing the chemical as a matter of everyday activity. This kind of crisis is survivable especially if company leadership is immediately "on top of it" and offers a sincere apology.

Hospitalization of two or three individuals might also constitute a Level 1 crisis under Klann's definitions. I would argue that a single hospitalization probably isn't even a Level 1 crisis, but presumably, two or three such injuries probably are.

Ethical breaches that rise to the notice of the newspaper or top management also constitute a Level 1 crisis. Even if the offending person is terminated, the company takes its lumps in the newspaper and it's back on track pretty soon. There is some public embarrassment, but the effects are transitory in nature, particularly if a worse crisis is prevented by leaders acting immediately to control effects and throttle future incidents.

Recall in 2008 that Fannie Mae, the financial institution, *used to* have a good reputation. Their involvement in the fiscal crisis of 2008 permanently tarnished their public image. Damage could have been limited to a Level 1 crisis, but they managed it poorly and needed zillions of public dollars to stay afloat.

General David Petraeus was famous for conceiving and garnering presidential support for the "surge" in the Iraq war and later became famous for being Central Intelligence Agency (CIA) director for 14 months. After his extramarital affair was exposed in late 2012, most people, including me, thought his career was over. But in Spring of 2013, after being out of sight for five months, he very painstakingly planned his own survival. He offered a sincere apology in a very public press conference of students and veterans, and he said he wanted to continue in public service. In fact, Phillip Ewing, who writes for *Politico,* recently (https://www.politico .com/story/2015/03/david-petraeus-plea-115723.html) said the following about the general after he decided to end the controversy quickly and decisively after a plea bargain agreement in which he admitted leaking sensitive information:

> "One example of Petraeus' lingering strength was the paucity of public criticism following his plea deal on Tuesday," said Ewing.
>
> "So although Petraeus almost certainly will not reenter government service in the final years of President Barack Obama's term, his camp believes that Tuesday's plea could set him on course to get into the strongest position possible for a future commander in chief."

One of Petraeus' staffers said, "I can't imagine a scenario where, if a president someday, no matter which party, called David Petraeus and said 'I have a job for you,' I can't imagine a scenario in which Petraeus says, 'nah I'm good.' It's not in his nature. It's not in his bones to ignore a call to serve."

In my opinion, General Petraeus' remarkable brand of determination and self-effacing behavior shows a masterful handling of a Level 1 crisis where Fannie Mae failed.

What about a Level 2 crisis?

In a Level 2 crisis, there is personal injury or fatality: In the case of the Tylenol tampering case, seven people died in the Chicago area, and the potential for more injuries remained for a few days until Johnson & Johnson's leaders could begin to manage the crisis. J & J "owned the problem" immediately, which Fannie Mae never did. Klann doesn't exactly say it this way, so I am interpreting his work, but it seems obvious that because public embarrassment and the potential for a tarnished image are also present in a Level 2 crisis, a Level 2 crisis includes the effects of a Level 1 crisis plus the need to *take immediate, public action by authentic leaders* to thwart even worse effects. This isn't a case where a sincere apology is enough, as it was for David Petraeus: An immediate and public counteroffensive is needed because the fallout occurs to many people and the finances of a huge corporate giant. We see that Johnson & Johnson's leaders did a good job at a tougher crisis than Fannie Mae had or even David Petraeus, and it emerged almost unscathed.

A Level 3 crisis can be survived with the right leadership

Here, the loss of property is significant, the likelihood of injury is high, and even fatalities are possible. In this "do-or-die" situation for an organization's leaders, public embarrassment is a given. Losses are moderate to significant, and the need to act is even more immediate.

As a worst-case scenario, the company's reputation may not be salvaged, even under strong and authentic leaders. Klann nominates Enron as the best example of a major player reduced to bankruptcy by massive unethical breaches.

I have an even worse example of a leader failing under duress: the matter of "pink slime."

In 2012, a company sold meat products to hundreds of schools for lunch, and they called it "lean, finely textured beef." Somehow, and probably from an ABC News story, the product got the nickname "pink slime," and after less than six months, the parent company, Beef Products, Inc., ended up laying off most of its workforce in the ensuing crisis of public trust. The company's leaders simply did not act quickly enough, or with enough determination and did not attempt to counteract the innuendo. Not even strong, authentic company leaders were able to salvage the institution from this kind of negative branding, and there is good evidence that the products they sold were perfectly good but somebody sabotaged the company with the detrimental name. Enron is gone, and some of its executives are serving prison terms; despite filing a lawsuit against ABC News for the destructive news story, Beef Products, Inc., will follow Enron into receivership. Both companies experienced a Level 3 crisis from which they did not emerge.

Here are two other case studies of leaders reacting to a Level 3 crisis. I am reminded of two local examples in the recent decade. One company did not survive and the other flourished, and both centered on leadership. In the first case, where the crisis essentially killed the company, The Dick Corporation, located in Pittsburgh, a large family-owned construction company apparently allowed or missed the use of some nonspecification nuts on a huge cantilevered steel beam where a new convention center was being built. The nuts stripped, possibly even a single nut, and a steel beam crashed down, with a resulting fatality (see https://www.osha.gov /pls/imis/accidentsearch.accident_detail?id=200540714).

The company struggled on for a few years after the crisis and found itself a financial savior to lead it out of near bankruptcy, but now, nobody sees it doing any work in the Pittsburgh area. Indeed, the company's former corporate headquarters is unoccupied. A few people have traced the collapse to the national financial crisis of the later 1990s, but most construction-savvy people attribute the final "straw that broke the camel's back" to the two years' worth of bad press associated with the fatality. Dick Corp never seemed to want to "own the crisis." They could have mitigated the damage to their reputation, but they weren't sorry until they got caught defending the indefensible.

In the second example, a small coal company experienced an unexpected crisis at a local mine. The mine's underground workers broke into a very old but unmapped mine that had been abandoned decades earlier. Nine miners were trapped in the now-famous Quecreek crisis, and after a week being underground, hope was rapidly slipping away.

The mine owners took a long-shot chance that a tiny local company, Yost Drilling, mostly known for drilling air shafts for local mines could help. Yost was a small company but highly experienced with big bits. On the day of the Quecreek incident, Yost was just about the only company in the United States with the experience and necessary equipment to drill a large enough hole to rescue the miners, and they volunteered to help. They decided to use a drill bit that weighed close to a ton to drill a 24-inch hole, using GPS coordinates, to the miners below.

In fact, in retrospect, witnesses said it was a calculated gamble to use the 24-inch bit since the rescue basket was 22 inches wide and the hole was only 24 inches. There was only an inch margin on each side. A small rock could have stalled the basket coming up.

But forging ahead bravely, Yost arrived and immediately began drilling. When that bit got permanently stuck, Yost's crew didn't panic. They knew exactly where to get a second and probably the only remaining bit in the United States. Yost crews drilled around the clock, and it was their hole that allowed the nine Quecreek miners to be rescued (Figure 10.4).

Yost was quickly bought by the huge Chesapeake Energy Corporation, not in small part because of the positive media coverage and distinct down-home brand of leadership it displayed during the nine-day crisis to rescue the Quecreek miners. In this Level 3 crisis, Yost's leaders acted boldly; they were fully competent and they didn't let setbacks to divert their attention from the crisis at hand. Leading boldly had positive effects on the company itself.

Figure 10.4 A miner emerging from the Quecreek Mine using the very narrow basket.

In fact, Klann agrees that crises can be opportunities. His 2003 book outlines four of these opportunities for leaders to emerge under crisis conditions, which I apply to the Yost example:

- Hardships allow individuals and organizations an opportunity to examine their own core values. Yost's employees personally knew dozens of miners and knew they had to act immediately.
- Crises bring out courage, honor, selflessness, and many other positive behaviors. Yost's leaders acted selflessly. They did not call attention to themselves.
- Handling a crisis promotes confidence and personal growth. Yost's leaders didn't see this at the time, of course, but their company's reputation spread worldwide as a direct result of their success.
- A crisis can create bonding and a keen sense of camaraderie. Yost was, and is, looked upon as the model driller in the southwest corner of Pennsylvania.

Having a crisis may end in further disaster for the organization, but not necessarily. Strong and authentic leadership—the kind displayed by Yost employees when lives were on the line—makes the difference whether the organization emerges and even prospers as a result. The take-home message is that the level of crisis isn't as important to survival as having well-prepared, authentic leaders who choose to step up.

From here, we will explore two models for noncrisis leadership, namely, *Servant Leadership* (2002) of Greenleaf et al. and Collins' *Level 5 Leadership* (2001). After these, we'll examine two models for crisis-based leadership, including Kolditz's *In Extremis Leadership* (2007) and then *Leadership in Dangerous Situations* by Sweeney, Matthews, and Lester (2011).

Noncrisis leadership model no. 1: Servant leadership

I remember my first exposure to the idea of servant leadership. Col. Dave Miller at Virginia Tech, a family friend, explained that it is the model used by their corps of cadets: the idea that a leader is first a servant to his or her people. I thought surely the order is reversed. How can a leader be beneath the servant? But it became clear in my discussions with Col. Miller that a leader in this model only truly leads when the needs of his or her people are served first. Those needs include not only their material immediate needs of food and shelter but also emotional support, enthusiasm, and much more. The idea of servant leadership borders on the religious, a conclusion shared by many, but it is much more.

Robert Greenleaf is credited with formalizing the ancient concepts about organizational relationships in his first book from 1977 titled, *Servant Leadership: A Journey Into the Nature of Legitimate Power and Greatness.* People took notice, including some of the best-known writers on organizational behavior. Stephen Covey (1989), for example, in writing his blockbuster work *The 7 Habits of Highly Effective People*, credits a person's character, not simply personality, but the fundamental nature of putting people first, as most influential in stimulating positive change.

Covey met Greenleaf as a student and claimed to be permanently changed by a presentation he heard then. Now, when we hear things like "putting people first" or "win–win situation," we are hearing it from the work Covey began. Indeed, the whole thought behind "our people are our most important resource" is something we've all heard many times, and it comes directly from Covey and his reliance upon the servant leader model. Managers who create a win–win situation are directly beholden to Covey's (1989) Habit No. 4.

Another influential writer who believes in the importance of the servant leader model is Ken Blanchard, author of the well-known *One Minute Manager* (Blanchard and Johnson, 2003). About servant leadership, Blanchard and Covey agree: Putting people first means finding out what they need to be successful, and this is something a manager is not called upon to do. In an essay written in 1991 in the *Blanchard Management Report* titled "Servant Leadership," Blanchard writes in the following excerpt from that essay:

> The traditional way of managing people is to direct, control and supervise their activities and to play the role of judge, critic and evaluator of their efforts. In a traditional organization, managers are thought of as responsible and their people are taught to be responsible to their boss.
>
> We're finding that kind of leadership isn't as effective as it once was. Today when people see you as a judge and critic, they spend most of their time trying to please you rather than to accomplish the organization's goals and move in the direction of the desired vision. "Boss watching" becomes a popular sport and people get promoted on their upward influencing skills. That role doesn't do much for accomplishing a clear vision. All people try to do is protect themselves rather than help move the organization in its desired direction.
>
> What do managers need to become servant leaders? The biggest thing they need is to get their ego out of the way.

Is servant leadership really a religion in disguise or religious in some way? The answer that Greenleaf, Covey, and Blanchard would probably give us is "no," but it's still close, in my view. Kent Keith's book, *The Case for Servant Leadership* (1998) has the best explanation I have seen so far for the close and very useful proximity of servant leadership to religion but shows that servant leadership exists without being a religion or having religious trappings or ritual. The universality of the desire to lead based on the desire to serve is far reaching, indeed, and captures basic elements of faith across the world. Maybe that's why it has such appeal as a model for organizational leadership.

In this same book, Keith summarizes the parallel between servant leadership and religious tenets from various formal religions across the world from, which I have sampled:

- All men are responsible for one another (The Jewish Talmud)
- Men must carry each others' burdens (Paul, first letter to the Galatians)
- If I employ others for my own purposes, I shall experience servitude. But if I use myself for the sake of others, I shall experience only lordliness (Bhuddist text, *Shantideva*)
- To realize the pain and suffering of others and to offer your hands in assistance, helping to alleviate their suffering, is Islam (Sufi Sheikh M.R. Bawa)

James Sipe and Don Frick are also proponents of servant leadership. In the 2009 revision of their book *Seven Pillars of Servant Leadership*, they provide an understandable structure for what a person of character must do to become a servant leader. I am actually only doing a passingly credible job on summarizing their main points in their book. I strongly recommend getting this book, especially if you are going on a cruise or someplace where you expect to get some real down time to read.

Pillar 1: A servant leader must become a person of character

The leader makes insightful and ethical decisions. The servant leader maintains integrity, demonstrates humility, and acknowledges that he or she is serving a higher purpose. The authors say there is no reason to progress to the remaining stations of servant leadership if the candidate cannot or will not become a person of character. That person who listens to his own conscience and pays attention to his own moral compass.

We have previously defined an "authentic leader" as acting on personal convictions and "staying on course" when shades of gray intrude. Being an authentic leader is roughly the same as being a person of character.

Serving a higher purpose is part and parcel of the successful safety leader or project manager engineer because people's lives are at stake. The higher purpose here is sending all of your employees home safe at the end of a hard shift.

Pillar 2: A servant leader puts people first

Here, the leader is inspired to serve before he or she leads. As counter-intuitive as it may seem at first, the servant leader makes sure that his or her peoples' needs are accommodated before his or her own. Napoleon and Frederick the Great were known for eating meals with their soldiers by standing last in line. When he was wounded at one point in the Civil War, Stonewall Jackson gave his own ambulance to an enlisted soldier whose wounds were worse. Not only are these ennobling stories, but also, these build unmatched loyalty. Neither Napoleon, Frederick the Great, nor Stonewall Jackson would tell these stories themselves; somebody close by noticed and told the stories later.

Here's a simple test that I bring up because I have seen it in action. You do something for your employees, even your family, that is truly enno-bling. For example, you used your annual bonus to support the local soup kitchen or food bank. This is something you didn't have to do but it made their lives easier or more secure. Be the first to tell no one. That's right. If you are eager to tell people around you that you've recently been enno-bling, you aren't ready yet for becoming a servant leader. You still haven't buried your ego; your own needs are still coming out on top.

Pillar 3: Servant leaders must become skilled communicators

Here is one of the most critical jobs a safety leader or young engineer is going to need. We are in the business of interfacing between the company's loss control/safety function and craft workers. As Sipe and Frick say, we'll need to be empathetic but persuasive. We'll need to invite feedback. We'll have to hone our people skills.

We have to communicate regularly up the line to upper management, as well as down the line to foreman and craft workers.

This was apparently one of Robert Greenleaf's most important con-cepts of leader development. He noted that listening first more clearly denotes a servant leader, as opposed to talking first. How many people do you want to have a conversation with—a real heart-to-heart—who are talkers first? Those people are pretty much off my list after the first day. Maybe I need to post a sign on my door that says, "If you want to talk with me, welcome. If you want to talk at me, please go next door."

Do you ask questions about people's lives? Do you ask about their families? Do you ask how their hunting trip went, or how their mom is doing in the nursing home, or do you get right down to business? Being a skilled communicator means drawing people out to talk about their con-cerns so you can work on solutions cooperatively.

Do you ask more than you tell? I remember Francis Hesselbein quot-ing Peter Drucker recently something like this: Judge the quality of your conversation, in part, by counting the number of periods and the number of question marks. Remember again about burying the ego: The skilled communicator is probing and rarely pontificating. Which person would you rather be around?

Pillar 4: The servant leader is a compassionate collaborator

This is the person who thanks team members for their contributions and who tactfully negotiates conflict among members. In a true collaboration, once again, we leave egos and calling attention to oneself at the door and build teams, respect diversity, and accept personal responsibility for decisions.

Frick and Sipe cite the contribution of the famous Harvard Negotiation Project authored by Roger Fisher in his award-winning book *Getting to Yes: Negotiating Agreement Without Giving In*, dating back to 1979. The book and its 2005 revision, *Beyond Reason: Using Emotions as You Negotiate*, suggest a few guidelines for arriving at a mutually beneficial agreement: Watson calls the five guidelines the Peace Rules:

1. Remain calm and be respectful.
2. Unto others as yourself, avoid hostility and blame.
3. Listen to understand; don't be the person who fills every gap in a conversation with words.
4. Expect success for the group and stay there until you get it.
5. Seek outside support if you need it. Sometimes, a third party has a much clearer understanding of the conflict and can explain differences better than group members can.

Pillar 5: The servant leader shows foresight

As Sipe and Frick say about Greenleaf, "foresight includes—but goes beyond—traditional planning activities to have a sense of the unknowable and to be able to foresee the unforeseeable."

Of course, in reality, it is impossible for safety pros or project engineers to foresee the future, but there are those special people who seem to be able to do it, and followers are in no short supply for these special leaders. They inspire confidence and exude competence.

Maybe it's intuition or a sixth sense, but we know when we're around people like this. But according to Sipe and Frick, the servant leader has an obligation to act in this way. They can "eyeball" the future using only gut feeling; they are the people we want to be with in the car whose driver is lost in the two lane roads of Iowa.

Sipe and Frick say that leaders who study the past, leaders who learn everything there is to know about a topic of potential conflict, leaders who then let things incubate for a while and don't jump to conclusions at the first opportunity, leaders who are sensitive to breakthroughs—those leaders have prepared themselves to be "blessed with foresight." Not without doing a lot of homework, of course, but these leaders will inspire others because, most often, and by using these methods, they're right.

Surprised? I'm not. I've said for years that people make their own luck, and I'm convinced that this notion of "foresight" is more a matter of careful study. Call it luck. I call it preparation.

Pillar 6: A servant leader is a "systems thinker"

The servant leader is known for seeing things as an organic whole. As Greenleaf noted in the Sipe and Frick text, organizations can accommodate periodic disruption and conflict even when the whole of the organization is moving in the right direction anyway.

Systems analysis itself has been a key component of safety with its own branch; even systems engineering as a field has developed in the past couple of decades as a way to model events and causes and provide feedback loops for constant adjustments, as necessary. We can apply the definition of systems analysis here.

The servant leader makes himself or herself as comfortable as possible with the sheer complexity of systems or at least works at it when the whole is overwhelmingly big. Listening to GE's Jack Welch in class as we have done, I am constantly amazed that this guy really did seem to have a handle on all of the dark corners of his entire organization. He knew where he wanted to be, and for the most part, he had a plan to get there. That's seeing the "organic whole" of the organization.

This is probably harder for a trained engineer to do than a trained safety professional because engineers delight in the minutia of solutions and less often in their effects, and this seems true especially of entry-level engineers. I am reminded of the vicissitudes of the human species called "engineers" periodically by my friend, Ralinda Miller, who is a professional geologist and editor of complex technical publications. She is surrounded by competent engineers doing environmental remediation, but sometimes, they simply think she should automatically know—and consequently the public should know—what they mean when they use a table or formula without explanation. Sometimes, they are quick to anger when she asks them to explain their ideas more fully in the publications. She says tongue in cheek and in her own good natured way, "Dilbert lives" after she explains the need for the bigger picture to the engineers.

The servant leader who takes time to remind is or her subordinates about the macro view—why we're here, what our product is, and who our customers are—is well served when the subordinates get into trouble. Providing the macro view smoothes things over. Ask Ms. Miller.

Pillar 7: Servant leaders are granted moral authority only by their subordinates

The servant leader does not demand respect and authority as much as he or she acts with respect, civility, and humility and earns the moral authority. Greenleaf says it best in Sipe and Frick:

> ...the only authority deserving one's allegiance is that which is freely and knowingly granted by the led to the leader in response to, and in proportion to, *the clearly evident servant status of the leader* [emphasis added]. Those who choose to follow this principal will not casually accept the authority of existing institutions. Rather, they will freely respond only to individuals who are chosen as leaders because they are [first] proven and trusted as servants.

Greenleaf is saying that authentic leaders act with moral congruency in their daily life, at work, and among family. They earn moral authority and they do not demand it. They earn their status first as servant before they act as leader because of this congruency.

In summary, Robert Greenleaf first recognized quite brilliantly that person-to-person caring was something we used to do instinctively. But now, because governments and private institutions have grown so dramatically in complexity and size, most caring is institutional and, therefore, hollow. In the absence of a clear and shared purpose, institutions don't demand, much less earn, moral authority; institutions can't think globally; institutions can't show foresight based on planning and experience. Most assuredly, institutions can't collaborate, and even more assuredly, institutions don't put people first. Yet this is the essence of the business in which we find ourselves engaged as safety professionals and engineers. I admit that maybe it's too close to a religious experience or thinking that suggests that we put people before the institution that pays our wages, but it's still a very useful idea of how to become a leader and not just in name only. They earn it.

Does servant leadership imply weakness and consequently find itself incongruent with military leadership? On the contrary. We will soon evaluate Thomas Kolditz's *In Extremis Leadership* (2007) model to see what we can extract for our safety professionals and engineers.

Here's what he says about servant leadership in the *HBR* from back in 2009:

> In your own development as a leader, have you found value in putting other people first? Did it seem out of place in competitive, results-oriented businesses? Did it powerfully influence people, or did it merely suggest weakness? And have you had role models in business who you see as effective because of their servant leader orientation?
>
> In many business environs it is difficult to inculcate a value set that makes leaders servants to their followers. In contrast, leaders who have operated in the crucibles common to military and other dangerous public service occupations tend to hold such values. Tie selflessness with the adaptive capacity, innovation, and flexibility demanded by dangerous contexts, and one can see the value of military leadership as a model for leaders in the private sector. (*HBR*, February 6, 2009)

There's just so much good to say about "servant leadership," so in closing, here's what Kent Keith says. Recall that a couple of people I asked to refer important books for young professionals' reading lists mentioned his book, *The Case for Servant Leadership*:

Kent Keith's Paradoxical Commandments of Servant Leadership

- People are illogical, unreasonable, and self-centered. Love them anyway.
- If you do good, people will accuse you of selfish ulterior motives. Do good anyway.
- If you are successful, you will win false friends and true enemies. Succeed anyway.
- The good you do today will be forgotten tomorrow. Do good anyway.
- Honesty and frankness make you vulnerable. Be honest and frank anyway.
- The biggest men and women with the biggest ideas can be shot down by the smallest men and women with the smallest minds. Think big anyway.
- People favor underdogs but follow only top dogs. Fight for a few underdogs anyway.
- What you spend years building may be destroyed overnight. Build anyway.

- People really need help but may attack you if you do help them. Help people anyway.
- Give the world the best you have and you'll get kicked in the teeth. Give the world the best you have anyway.

Here is my take on Keith's Commandments.

- The very best leaders I have ever known will press on even when the big boss isn't looking or doesn't care. That's character.
- The very best leaders continue to do the good but unnoticed work even when their supervisors seem to criticize their every movement; that's seeing the system is whole.
- Authentic leaders will get down in the dumps in private but show only a tough exterior to their direct reports. No matter what may come. That's putting people first.
- And moreover, those same authentic leaders make an active, conscious choice to act that way. Nothing about them leading was ever ordained or given or automatic. They chose to lead; they chose to act consistent with their own core values. That suggests that moral authority is granted by subordinates.

I will freely admit that there is no data-based foundation for servant leadership or Keith's commandments. It does have a religious kind of flavor, but caring and serving are central to the business we are in.

Robert Greenleaf and a lot of theorists and researchers think that people come first regardless, and I join them. Highly effective and renowned organizations have sprung from these fundamental truths about putting people first. Indeed, the most effective armies and the most effective organizations, as we'll soon see, have put these principles to work. It just makes sense to me.

Don't we want to be the safety professional who pays just a bit of extra attention on the new guy as he learns his job because it's about being servant first? Don't we want to be the engineer who visits her injured employee in the hospital? Don't we want the best possible safety program whether somebody puts our name on a plaque or not? I hope we do.

And I have to conclude this section with a personal note. The idea of leader being servant first, putting people truly first, and of a boss who leaves his ego at the door has changed my life, and the change is ongoing. I see the servant leader now in not only the boardroom but also the family, where the parent leads second by serving first. I see the servant leader as the best kind of community volunteer. I see the servant leader as being the best kind of professor.

Noncrisis leadership model no. 2: Level 5 leadership

Without a doubt, Jim Collins will be the best known and most prolific writer I discuss among leadership models that warrant our attention. His ideas resonate with me because of their simplicity and clarity. These lessons can easily translate to all of our fields and all of our chains of command. This stuff just makes good sense to me. Allow me to share it so that you can see the simplicity and clarity for yourself.

As a professor at Stanford, his first major work was released in 1997, *Built to Last: Successful Habits of Visionary Companies*, and it stayed on *Business Week's* top seller list for six years (Collins and Porras, 1997). His next real blockbuster book was *Good to Great: Why Some Companies Make the Leap and Others Don't*. It was published in 2001 and remains a staple of leadership studies worldwide. Collins has another book out, too, *Great by Choice*, released in 2011 (Collins and Hansen, 2011), but the jury is still out as to whether this book will be as influential as the others, or whether this is recycled or updated material.

Collectively, these books have been translated into dozens of languages. *Built to Last* and *Good to Great* represent the core of what has become known as Level 5 leadership. There are probably 500 YouTube videos on Collins, his work, or commentaries on his ideas, and typically, they are good because they are easily digested and they make logical sense.

When *Good to Great* came out, Collins wrote in the January 2001 *HBR* a fairly complete description of the methods and findings that support Level 5 leadership, from which I draw the following summary. In 1996, Collins began his basic research into why some companies thrive and some fail. It wasn't long until he noticed that some special companies actually sustained superior performance despite market downturns and changes in CEO. It was more than just these influences on the very best companies that distinguished them as above just good companies—they were great companies.

Collins started his research with 1435 companies from the Fortune 500 and examined data from 1965 until 1995. He used data from the University of Chicago for research in Security Prices and eliminated poor-performing companies. The research team also performed 87 interviews with corporate executives and collected over 6000 articles plus internal strategy documents. He ended up with only 11 companies, and these had average stock returns of 6.9 times the average stock market price. Collins also matched his top 11 companies as closely as possible with 11 comparison companies—a control group, so to speak. The controls were in similar industries with similar workforce sizes operating in the same types of markets and had similar geographical presence.

Collins points out in the *HBR* article that during that same period, Jack Welch's GE averaged only 2.8 times the average stock market price during his tenure there, making these good-to-great companies truly outstanding. Those good-to-great companies also sustained their performance for at least 15 years when compared against their controls. This work represents quasi-experimental research where true control groups and subject randomization is not possible, but group-matching and quasi control groups do exist. It isn't gold-plated research but it's pretty darned good.

Collins did make some mistakes, but only in retrospect. For example, he included Fannie Mae (the Federal National Mortgage Association) among his top 11 companies. As we all know, the mortgage giant was intimately involved in issuing or allowing the issuance of risky mortgages in the 1990s, leading to failure of the company in 2008 (seven years after Collins concluded his research and wrote *Good to Great*) and a federal bailout of over $200 billion. I wonder who was the Level 5 leader there and if Collins feels the same about Fannie Mae today. In fact, I would not be surprised if Collins still backs Fannie Mae showing how a company once down on its luck and laid low in the media can reinvent itself and recover. That's what Fannie Mae is actually doing.

Furthermore, Collins bombed in his promotion of Circuit City in his top 11 companies. While Circuit City outpaced its market competitors by 18.5 times between 1982 and 1997, it filed for bankruptcy in 2008, the same year Fannie Mae failed. Circuit City liquidated its inventory and closed its last store in 2009. Despite these missed calls by Collins, the mass of his work survives this minor storm and is widely embraced today among organizational scientists.

So what is Level 5 leadership? The Collins summary represents characteristic empirical findings that distinguish truly great leaders from just good leaders. And it isn't those huge, larger than life people as he says, such as Lee Iacocca, who make consistent headlines as Level 5 leaders; usually, they aren't. Curiously, it's the driven but humble leaders, the iron-willed leaders often working in the shadows who are the Level 5 leaders in Collins' top 11 companies.

Jim Collins is a widely acclaimed writer on the topic of how industries rise and fall, that is, what makes one company succeed when a similar company under similar economic conditions fails. Collins' books stretch across two full decades, and among the best known of them is *Good to Great: Why Some Companies Make the Leap...and Others Don't* (Harper Business, 2001). In *Good to Great*, he considers a distinct hierarchy of leadership, which I have described in brief, as follows.

Level 1: Highly capable leader

This executive or manager is already a competent leader and applies considerable talents and skills to address company or organizational concerns. But there is something missing, something that prevents the Level 1 leader from being more effective as a leader.

This leader isn't willing to take chances. He or she is good at bean counting, but not so good at setting up a real vision. In my work life, I have worked for about eight bosses: The average ones could stay on budget and hold effective meetings, but at the end of their tenure as boss, we had no long-term goals, much less made effort to achieve them, and they were ultimately forgettable. In one case, a chairperson went to a lot of trouble to draw up a strategic plan; it took eight months to prepare and deliver to his boss. Afterward, we faculty realized it had all been for show, a classic "Look at me" experience and worth nothing. The guy was highly capable but had no overriding vision.

Level 2: Contributing team member

This manager or executive is already highly capable. This person works effectively in groups in addition to the predictably good basic work habits.

I like working with people like this; we get the job done, and maybe argue a bit about schedules or minor things, but this person whacks out the work and gets it done. Give this person a job, and it will be completed.

But that's all that will happen. What about vision? Tomorrow? The really big ideas? The Level 2 team member is effective only at a low level.

Level 3: Competent manager

Building upon the characteristics enjoyed by both of the lower levels, this person organizes and links people and tasks working toward predetermined objectives. Once again, something is missing here to prevent the competent manager from working at the highest level: What's missing is vision.

Level 4: Effective leader

Combining the talents of the three previous levels, this person adds "compelling vision" and demands strict performance standards. Most organizations would be glad to have senior executives working at this level, but yet something is missing to prevent the manager from working even higher. This time, it's modesty and personal responsibility that are missing; that's right, two elements of the best leadership possible, something that costs zero but whose value is infinite.

Level 5: The Level 5 executive

The missing ingredients finally show up as personal humility plus strong professional will. Here, we find at the very top of Collins' successful companies compelling modesty and self-effacing personalities and not "look at me" people and headline-driven ego-maniacs. "When things go poorly, however," Collins says, "a true Level 5 leader will look in the mirror and blame themselves, taking full responsibility. The comparison CEOs often did just the opposite—they looked in the mirror to take credit for success, but out the window to assign blame for disappointing results (Collins, 2005).

In *Good to Great*, only 11 company executives displayed Level 5 leadership out of the hundreds and hundreds of potential executives. But I can hear you saying, "we waited all this time for personal humility and professional will? It doesn't seem like much to differentiate the competent leaders below Level 5."

But it is.

The Level 5 leader is a plow horse and not the show horse, to use Collins' apt analogy. The Level 5 leader is, however, often overlooked for internal promotion because he or she is self-effacing. Collins is not clear how this particular paradox is addressed; that is, how do the Level 5 leaders work through the mess and make it to the top? He does not say clearly.

More optimistic is the notion that Level 5 leaders can be made, and they are not just born. Collins suggests that Level 5 leaders are all around us if we just know what to look for. We'd look for competence and diligence, yes, but we'd also bypass the "see me lead my people" leaders in favor of those who do good competent work with humility and personal responsibility.

We know that Level 5 leaders are simple and self-effacing, fairly common people, not ego-driven and not the dazzling, wing-tip shoe guys that we might otherwise expect. In fact, Collins says he did expect to find showy types that at Level 5, but he says he did not. It's kind of underwhelming until we realize there actually is more to the story of good-to-great companies.

The shift from merely good doesn't happen without Level 5 leaders, who must also implement six other characteristics of these exceptional companies.

Let's talk about the six important characteristics of these companies that are created by Level 5 leaders. This represents the real meat of his findings.

1. *Get the right people on the bus and the wrong people off the bus*

   Among the most persistent companies, Collins' research team noticed that the best leaders did one immediate thing when they took over. Using Collins' metaphor, they got the right people on the bus (into the company's leadership structure) and ushered them to the right seats (positions of leadership where their talents could be used to best advantage), and they got the wrong people off the bus (terminated).

Collins doesn't say how to assess who are the "right" and "wrong" people. I can propose, then, that these can be fairly quickly and reliably identified by performance reviews, history of peer assessments, personality inventories such as we discussed in Chapter 5 (Core Values), and just doing your MBWA and listening carefully. To me, the right people are those who try hard, work in teams, contribute regularly, have forward-thinking ideas, and want to achieve. The wrong people are the classic "see me do my really difficult job" people and those who regularly shun responsibility or those who whine about everything. The truth is, these people really aren't that hard to spot.

Collins recognized that it isn't just people that make a company great, it's the *right* people in the senior executive team that can take a problem and turn it around. The wrong people can't turn around problems, and letting them hang on—making excuses that they're really nice people but they just can't achieve—deflates the morale of everyone else. Before any other single action, the truly great companies have executives who constantly and aggressively refresh staff. First is "the who" and "then the what," he says.

Collins says his research team thought that the truly great companies would begin with vision and strategy first and then be dominated by personality, but the opposite occurred. Vision and strategy came after assembling teams, and personality rarely played a big part. This is counterintuitive, yes, but it is also exciting.

In his book, and thankfully so, Collins directly addresses difficulties faced by certain academic institutions with tenure commitments to faculty and government institutions where unions and momentum make it almost impossible to get the wrong people off the bus. Never fear, he says clearly in the book, it just takes longer. A Level 5 leader keeps his or her eye on the prize, hiring only the right people and the strongest minds available, then that leader starts "gradually creating an environment [of enthusiasm, hard work, and accountability] where the wrong people increasingly feel uncomfortable and either retire or go elsewhere."

To that, I'd add my two corollaries for Level 5 leaders: First, don't say you have true merit pay for exemplary work when you hand it out like welfare for nonproducing employees, and second, don't provide high-profile recognition for self-promoters who have assembled three-ring binders on their "selfless service." In academics and in industry, the transformation to "great" may take longer, but the Level 5 leader, if we believe Collins, never loses sight of replacing deadwood with top-shelf performers at every opportunity.

2. *Collins' hedgehog concept*

In the second principle, what Collins calls his "hedgehog concept," he says that the fox is always using cunning and a myriad

are reported. Within weeks came demands for ways to provide in-mine shelter, methods to install different breathing devices, and a dozen other initiatives made by well-intentioned people in hopes things got better fast. Those people, plus the legislature, plus the media, wanted it all done *now*.

Dean understood that control and a measured pace, especially during a crisis, were more important than making changes for change sake, I knew, for example, that in addition to being a part-time farm operator, he took two years to build his own house. Slow but deliberate wins the race.

"You can't walk over and flip the switch and change mine safety in a year," said Jim who spent almost a year as West Virginia's mine safety chief after the Sago Mine explosion. "The negative is, it's not happening fast enough," wrote Tim Huber in a *New York Times* article about Dean's work with the state and the post-Sago regulatory climate in 2007.

Dean worked steadily on each initiative, first as interim director and then chair of a task force to continue the work. In the intervening years, a host of changes have been made: Refuge shelters are now required underground and improved breathing devices are available, plus enhanced miner training for escape. Dean helped write a bill called the Mine Improvement and New Emergency Response Act signed by President Obama in June of 2007.

It is important to note that each of these actions came with cooperation from the coal industry and from the United Mine Workers. And those who know Jim recognize that he still believes in his work to improve mine safety. He is the first one at the office in the morning and one of the last ones to leave.

"Getting legislation in place 'just took a while,' maybe a little bit longer than I wanted, but we succeeded in making huge improvements. I never doubted that we'd get it done" said Dean as he walked back to his tractor.

4. *Develop a culture of discipline*

Assuming that we have already got the right people on the organization's bus and ushered the wrong people off the bus, the next steps are easier. The goal is to create a culture of discipline without rigidity. That leads, says Collins, to setting up a roadmap of disciplined thought and disciplined action, and a lot of it comes from side-stepping (or purposely ignoring) unproductive bureaucracy.

With disciplined people, thought, and action, your organization can trim bureaucracy, rules, and artifice designed to provide accountability. Trimming the bureaucracy speeds up the flow of ideas and paper. The entire staff, labor force, and upper tier management are focused on "what we do best" and not much else.

Once you know what you do best, you also stop doing everything else, says Collins. You stop reallocating money to faltering pet projects; you stop funding unproductive product lines; you have a longer and longer "stop doing that" list and a shorter list of projects requiring full concentration.

Case study: failing to learn a lesson about product quality

I once worked for a helmet manufacturer that just couldn't get the quality right on its face shield to seal the front opening and block noise and wind from entering. The plastic ratchet mechanism has been designed such that the darned thing never pushed the shield tightly enough against the helmet to block the wind and noise. Management seemed much more committed to keeping the face shield's designer employed, and not the quality of the design itself, which should have been canned. We weren't disciplined enough to keep the focus on customers who demanded a lot more for their $400 helmet.

Finally, a big international distributor sent back an entire container load of these expensive helmets from Italy because of customer complaints about the face shield fit and the 99-cent plastic closure mechanism. Did that cause enough management consternation to promote change? Nope. We crushed the entire container full of helmets. We remained focused on short-term advantages of not having to pay to redesign the problem, we continued to bow at the feet of the anointed designer. We should have been disciplined enough to focus on customer satisfaction. We did not have a climate where truth could be heard above the din of self-congratulatory buzz by the design staff.

5. *The flywheel effect: building up momentum and ignoring the quick fix*

Collins' great companies such as Kimberly-Clark and Wells-Fargo worked slowly and surely and didn't get distracted by fancy spendy-technologies or somebody else's "next big thing." All of Collins' "great" companies had at least three times, and often ten times, the amount of market returns as their comparison companies did.

Rather, the great companies were like a flywheel, to use Collins' analogy. A flywheel starts slowly and never changes direction. It builds momentum, and if the analogy works, there is some point along the path of gathering speed that a breakthrough happens: The engine runs on its own, and the company owns its part of the market. The flywheel causes the engine—and the company—to roll through a spark misfire or a market downturn. That's the beauty of the analogy (see Figure 10.5).

Collins says the great companies are not distracted by vagaries and fluff, they are staffed by disciplined people, and they build momentum like heavy flywheels that never change direction. Finally, bang! After maybe years of work, the engine goes from merely rotating mass to a useful mechanical tool. And since we know that Newton's first law says an object set in motion (the flywheel) tends to stay in motion unless acted upon by an outside force such as friction (competing markets), the flywheel analogy makes sense for understanding organizational behavior. The faster the flywheel, the less it is to be affected by the outside force.

My own nomination for a great company that best meets Collins' flywheel rule is Apple. It was given last rites in the late 1980s and into the 1990s. And it had some duds along the way. Do you remember the IMac G4 Cube? It was a market flop, and there were a few more. But there were fewer flops each year.

Eventually, Apple's cofounder, Steve Jobs, was edged out by his own board, until it became clear that only Jobs himself had the vision—and courage—to be disciplined, flywheel-like, and look only to the long run for success.

Figure 10.5 Note the heavy flywheels on this stationary engine being repaired. On a field trip, my engineering students and I explored the advantages of low-tech designs.

Jobs' strong personal will and fierce dedication to what was at first a very thin market overcame even the top players like IBM and Microsoft to become the most valuable company of all time in 2012. All of what we take for granted about graphic user interfaces, using the mouse instead of key strokes to guide the computer cursor or iPad operation, and above all, minimalist and elegant design, all flew in the face of the heavyweights of the day.

Jobs never gave up turning the flywheel and gaining momentum slowly, from being an unknown software maker to making the ungainly Apple 4e computer to the now iconic iPod, to the much emulated iPhone, to the trendsetting and 100-million-selling iPad. The iWatch is evidence that once again; Apple beat the market in a new product about which there is controversy.

But it's happened before, and Apple succeeded before. Starting with one foot in the grave in 1985, it methodically stayed disciplined and looked only to the long haul ahead. But then after their breakthrough products, they pretty much own the market today for personal electronic devices.

6. *Use technology accelerators, but don't depend solely upon them*

In Collins' view, technology is not the instigator but merely the accelerator of great companies, just the same as it is the decelerator or the demise of many failing companies. His great companies valued and used innovative technologies. This was exemplified by companies such as Nucor using thin slabs (not 18 inches, like the old mills), continuous casting (not one-pour methods like old mills), and electric arcs (unlike coal-fired furnaces of old mills). Yet Nucor's CEO, Ken Iverson, didn't even count technology among his top five or six prized innovations when he was interviewed by Collins' researchers. Those were reserved for slim bureaucracy, thin layers of staff, and consistent philosophy about our customer and market. They used technology as a tool, not a crutch.

Buying new technology to keep up with the Jonses ignores the maxim that technology is only secondary to a *firm vision* and *dogged determination* to succeed in a vertical market. Technology by itself does not matter much at all unless it fundamentally and permanently improved the company's goals.

Fancy technologies are often showcased as the answer to plant safety. One of my students works for a coal mine that publically announced that it invested in "leaky feeder" (electronic "people-finder") communication technologies, but the same company allowed miners to work under an unsupported mine roof, a well-known cause of injuries and fatalities over the years. Failure to heed that fundamental safety rule of never working under an unsupported mine roof is a sure way to decelerate safety progress despite heavy investments in expensive and sometimes untested electronic people finders. I want to say to them: "Get your priorities right: Invest in technology after you are doing everything else right."

Technology should come second to the big things, the more important things like thinning the layers of bureaucracy, hiring only the best people and getting rid of the dead wood, adhering to the hedgehog concept, and so forth. Buying technology for its own sake—without a firm promise of fundamentally improving worker safety or product quality—is a *decelerator* of progress.

Summary of Collins' concepts and their use in safety and engineering

Jim Collins remains at the helm of the management and organizational behavior vanguard in the United States and has been for over a decade. His research involves carefully measuring market valuation of great companies against potentially great companies that are matched to their particular market, size, global status, and geography. His work is not psychosocial, and therefore, it is much more like quasi-experinemtal work than truly experimental work or hypothesis testing in a controlled laboratory.

Quasi-experimental work is more difficult in many ways simply because of the lack of available and tight controls, and sometimes, there are errors after-the-fact. Indeed, I have pointed out earlier that Collins missed seeing two of his great companies failing, but in fairness, both of those, Fannie Mae and Circuit City, were five or six years from the brink of failure and were still producing well when Collins studied them.

Even so, I have far more praise for Collins than not. He has used rational means of analyzing organizations even if the methods are not experimental. He finds that truly great companies—not just good companies—have exceptional leadership at the top that is dominated by iron will and, yes, humility.

Humility: We know it when we see it, don't we? I vividly recall reading Collins in *Good to Great* that David Packard, cofounder of the mighty and world-dominating computer and electronics company Hewlett-Packard, made sure that his epitaph didn't say "famous CEO" or "inventor of many products." Packard's epitaph said "rancher." That's humility, and that's my kind of guy.

Putting the same lesson into action, a young safety professional or project engineer would then rely upon discussions and advice from 30-year veterans of the assembly line and not put "safety professional" on their doors on the day after hiring in. That's humility, too, and it's a valuable piece of advice.

Likewise, an iron-willed engineer does not take shortcuts. He or she knows that buying steel beams with welded-on fall arrest anchorages costs more upfront but prevent ad hoc adjustments or shoddy tie-offs later. Those are two fabulous lessons to take away as extensions of Collins' research.

What else? Great companies have the ability to face the tough day-to-day decisions while keeping their eyes on the long-term prize but never varying from believing in ultimate success. Collins called this the hedgehog concept, and it works for me. He also suggests that technology for its own sake not only costs money that is diverted from core product and market thrust, but it also actually speeds average organizations along the downward slope toward their ultimate demise.

He also suggested that purchasing the latest safety gadgets that wouldn't be *fundamentally improving* the very nature of hazard abatement is more "show than go" and should be avoided. You don't have to be a Luddite, but you don't have to jump on every technology bandwagon either.

Of all his advice, I probably best appreciate Collins' notion of *getting the right people on the bus*. It means *selecting* only qualified workers with good records and asking prospective employees about their record and their involvement in previous safety programs, for starters. Getting the right people on the bus also means *inculcating values* and beliefs about safety through orientation and prework training; it means periodic refresher training for *every person* regardless; it means pushing safety *integration* to the lowest-level, least-paid workers regardless, and getting bad actors terminated eventually and assuredly, regardless; it means not knee-jerking to new procedures when things go bad but adhering to what you know works. These are incorporating some of Collins' notions of what constitutes great and not just average safety and engineering functions.

Getting the wrong people off the bus goes along with the former advice. It's more difficult because only a committed leader will work hard enough to remove the deadwood. It's easier to let the duds stay on, but keeping them will poison the system eventually. Collins says in so many words: Those people have to go, and go *now*.

Noncrisis leadership: The contributions of Zohar, Barling, and Kelloway

There are other researchers in organizational behavior that a young professional on the way to corporate leadership positions ought to be familiar with. These researchers (and there are others, of course) represent a slice of work that will resonate at conference presentations and in professional literatures. In the first example, I present a researcher whose focus (the notion of "safety climate") came then faded away completely, but has returned to center stage in the last few years. Still, this work is pretty easy to understand and applicable to OSH and engineering work.

I will briefly present researchers and the concepts that have gained prominence in the current literature on OSH leadership. These concepts are as follows:

- Safety climate
- Transformational leadership
- The importance of collaboration

Dov Zohar has a special place in the history of the safety profession because he was the first to use the term *safety climate*, which has been morphed since its first use in 1980 into the more popular term *safety culture*. An active researcher for 40 years and working out of the Technion Institute of Technology in Israel, Zohar created a quantitative scale for measuring safety climate and has tested various interventions against it. Zohar is one of a handful of experimental psychologists who have used strict experimental methods (such as quasi-randomly selected experimental and control groups) in testing hypotheses.

Safety climate and leadership are related. The former, according to Zohar et al. (2007), reflects "shared socially verified assessments of the workplace, i.e., which behaviors are likely to be rewarded and supported" (Figure 10.6). In this presentation, Zohar says that the safety climate of an organization reflects how its leaders truly act and feel. This is what I called values-consistent behavior earlier in this book; Zohar agrees when he says, in part, that a safety climate scale should measure what leaders *do* and not merely what they *say*. For example, he measures how well:

Leaders create *culture* and an *operational framework* through:
• Daily verbal exchanges between leader and members. This is
a key source of social influence (concrete task issues)
• Symbolic content or sub-text, as perceived by the recipient, which identifies deeper culture-shaping messages. It shows a leader's
    1. True priorities among competing goals and demands
    2. Formal policies vs. informal recognition (discrepancies)
    3. Espoused vs. enacted values (openness vs. authority)
    4. Words vs. actions (e.g., empowerment vs. control)

Figure 10.6 According to Zohar, leaders develop a culture through active and symbolic messages to subordinates. (After Zohar, D., Livne, Y., Tenne, O., Admi, H., Donchin, Y., *Clin. Care Med.* 1312–1317, 2007.)

My supervisor
- Refuses to ignore safety rules when work falls behind schedule
- Is strict about working safely when we are tired or stressed

Senior management
- Quickly corrects any safety hazard (even if it's costly)
- Considers safety when setting production speed and schedules

If my interpretation of Zohar's work is accurate, his data show that a safety climate is improved when leaders simply do what they say. Of course, we have seen this earlier when we said that an authentic leader acts values-consistent. But safety climate is sometimes considered to be the same thing as safety culture and the distinction has spawned dozens of papers in the last five or six years as researchers attempt to determine if there are really any useful differences.

There will be more papers dissecting the difference between "climate" and "culture," but we can draw a distinction here. *Safety climate* is the set of beliefs and perceptions that exist at a point in time, like the temperature at 7:00 a.m. today. *Safety culture* is a subset of an organizations' larger culture and exists, in part, as shared perceptions about long term adequacy of policies and procedures. Culture is similar to the weather generally, not the temperature at a particular time.

If the difference between safety climate and safety culture seems confusing to you, it is confusing to me, too. Apparently it is confusing to a lot of people and has been the topic of a 2013 conference sponsored by the U.S. National Institute for Occupational Safety and Health (NIOSH) and the Center top Protect Worker Rights (CPWR). That conference polled participants about how they understood "climate" and "culture" and arrived at definitions that may be useful in straightening out a puzzling area (see, "Safety Culture and Climate in Construction: Bridging the Gap Between Research and Practice, June 11–12, 2013).

It is probably unfair to Zohar for me to attempt to summarize his prolific work, but I will try. In briefest form, a given safety climate can indeed be assessed empirically by using scales he has published over the years—a leader doesn't have to guess whether his methods are having the desired effect. Second, a leader who knows his own motivations and values and acts in accordance with them will improve the safety climate of his or her organization. This is entirely consistent with what we have discovered in the findings of other organizational researchers.

Julian Barling is also an advocate of the application of empirical methods to the science of understanding leadership and leader development. Barling is the Borden Chair of Leadership and Queens Research Chair at the Queen's School of Business and has been an endowed professor for a long time. Although he has recently done some research on workplace violence, he is best known for his work on "transformational leadership," a term like Zohar's safety climate that is widely used and discussed today among organizational and behavioral psychologists. A transformational leader is most interested in his or her subordinates' welfare; they "elevate the interests of their employees; they generate acceptance for the mission of the group; they stir their employees to look beyond their own self-interests for the good of the group" (Bass, 1990, cited in Kelloway and Barling, 2000).

Kelloway and Barling cite extensive work that suggests a strong connection between transformational leaders and the work performance of their employees. These dependent variables cited in their 2000 paper include affective commitment that is raised by transformational leaders, improved sense of fairness observed by subordinates with transformational leaders, trust in the transformational leader (to advance the mission of the organization for the good of the group), and lower levels of job and role stress under transformational leadership.

So far, so good. But can a person be taught to be a transformational leader? One study was characterized, like Zohar's work, by strict experimental methods, including random assignment of test subjects to either experimental or control groups. Bank managers in a large Canadian bank were trained (or not trained in the control group) in a one-day workshop and four individualized counseling sessions (Barling et al., 1996). Subordinates reported significantly more positive perceptions of trained transformational leaders than in the control groups. In addition, trained leaders reported more commitment to organizational goals after training than did the control groups' bank managers.

Barling's research suggests that transformational leaders do what is right and not just what is cost-effective or expedient. In addition, transformational leaders provide intellectual stimulation and urge subordinates to help solve the organizational problems in new ways.

Reflecting his conviction that leadership can be studied and taught by using an evidence-based approach, Barling has published a book titled *The Science of Leadership: Lessons From Research for Organizational Leaders* (Oxford University Press, 2014). I recommend this book for anybody interested in transformational leadership. It is deceptively easy to read (what I'd call an airplane book because you can read it on the flight out and finish it on the return flight) but does not cut corners citing Barling and his colleagues' many years of organizational research on leadership. Make no mistake, this is a book based on the application of careful empirical methods to the study of leadership and leader development.

Kevin Kelloway is the Canada Research Chair in Occupational Health Psychology at Saint Mary's University in Canada. He has been a consistent research partner with Julian Barling on characterizing the attributes of transformational leaders. However, in recent years, his research focus has turned to workplace violence; he is lead author (with Barling and NIOSH's Joseph Hurrell, Jr.) of the book *A Handbook of Workplace Violence* (Kelloway, 2006). According to the publisher (Sage, 2006), these researchers summarize theoretical perspectives on violence and aggression and then discuss sources of workplace violence such as emotional abuse or workplace bullying.

Morten Hansen is a management professor at the University of California, Berkeley's School of Information, and in an earlier career, he was a professor at the Harvard Business School. He has written a book that merits the attention of upper management in just about all American industry, but particularly safety professionals and engineers entering the workforce. So when I discovered Hansen's book, *Collaboration: How Leaders Avoid the Traps and Create Unity and Reap Big Results* (Harvard Business Review Press, 2006), I was a little put off by the title offering "Big Results." Was this just a come-on to buy the book?

Apparently, there is really something important here for young professionals in OSH or engineering. Why? Because these particular leaders are charged with *cooperative interactions*, and probably more of them, among departments and levels of the organization than any other. I used to say in class that the safety professional and project engineer would soon have the largest number of contacts in his or her phone than anybody else in the plant. Hansen and I apparently agree.

I have further concluded that the Hansen book does have something to offer project engineers who must collaborate extensively. They coordinate with trade unions, a variety of skilled labor, community counsels such as zoning, regulatory agencies (both local and national), contractors and subcontractors, attorneys, land owners, plus many others. Similarly, safety professionals collaborate and coordinate with sales, operations, plant security, research and development, maintenance, and many more departments. Hansen has some surprising findings, and his book is surely worth reading.

Hansen gathered data on large companies such as Hewlett-Packard, Procter & Gamble, Apple, BP, plus a hundred others and determined that, sometimes, no collaboration is better than poor collaboration. That happens when there is "negative return on investment" that comes from lengthy hassles of working with other people and other departments who don't want to help you.

But what about safety professionals and project engineers, who must, by definition, serve the mission of preserving and protecting the people, property, and efficacy of the entire organization—they coordinate with everybody, don't they? Can they learn to avoid collaboration conflicts and still get results? The answer is yes.

Here are some tips offered by Hansen in his book.

- Sometimes people don't collaborate with others if they didn't discover the idea in their own department. This is Hansen's "not invented here" syndrome, which causes friction and prevents collaboration. It can be addressed by recognition of the fundamental problem.
- What happens when people have good ideas and don't share them with others who could benefit in support of the organization's mission? They "hoard," according to Hansen; they are in competition with their own coworkers. A good leader sees the hoarding and attempts to overcome it through a frank and honest discussion of the benefits and costs of hoarding.
- Hansen says that in larger companies, people spend up to 25 percent of their time just trying to find the right people to ask a question. This is a difficult conundrum for safety and engineering leaders: Hansen says close physical proximity helps, as do databases and knowledge management systems. Hansen says, "don't wait." Ask for help getting ideas out and open for discussion.
- It is difficult to transfer or share what Hansen calls "tacit knowledge" as opposed to "explicit knowledge." This is because tacit knowledge (maintaining an old engine, for example) is intuitive and subjective and might not come from a book. Collaborating and sharing tacit knowledge are more difficult for units with weak relational ties, constraining the flow. A strong leader will recognize the informational choke points and attempt to improve informational flow by encouraging and strengthening ties between units.

Crisis leadership model no. 1: "In Extremis" leadership

Is noncrisis leadership different from what a leader does in the event of a crisis? Probably so, at least according to available research which I'll present in the following sections. Subordinates simply expect more from their leaders when the building is burning or when a bad decision means somebody may die, to paraphrase Jim Collins again. Let's examine what a safety professional or engineer can do to become the best leader possible under the most extreme conditions.

Discussions of crisis leadership didn't start with West Point's Tom Kolditz, formerly the chair of the Behavioral Science and Leadership Department. But Kolditz is probably the single individual most often associated with the term *crisis leadership*. There are lots of books on financial and strategic crisis leadership, and yes, most of them are words of wisdom or sage advice rather than any real attempt to collect data and draw inferences based on empirical findings. Kolditz used content analysis on interviews with experts in preference to merely sage advice. This is real research and I was lucky enough to learn about it first hand by talking to him directly.

I met Tom Kolditz during a series of interviews I was doing for background for this book. The day I talked to him, having read his book, *In Extremis Leadership: Leading as if Your Life Depended on It,* he was holding a garage sale at his residence on the West Point campus. He was preparing for a new position at Yale University as a professor in the School of Management, and moving his household goods and family to New Haven, Connecticut. He was buzzing around the office talking to the person who replaced him as head of the Behavioral Science and Leadership Department, Col. Bernie Banks whom we have encountered before. I was pleased he took time for me.

As bit of small talk ensued during the initial part of our discussion, I must have hit a sensitive note when I told him I was interested in the similarities between what a young platoon officer does as a leader of people under difficult conditions and a young safety professional or engineer who does precisely the same thing. The similarity is that both young people are professionals who lead in the usual way, motivating and directing groups of followers toward a common purpose, but they sometimes must do so under conditions where the followers believe that leader behavior will directly influence their physical-well being or even survival. Those are what Kolditz calls *in extremis* conditions.

Kolditz was educated outside West Point in sociology, where he studied beliefs and attitudes among organizational leaders (see Figure 10.7). Kolditz says that his notion of leadership is pretty much what everybody else agrees that it is, with one exception. The general definition he uses, giving people purpose and direction by motivating and directing them, is fine for non-extremis conditions. But here he gets serious, and this is a very important point for young safety professionals and engineers to attend: extremis leadership is "conduct provided by an appointed or apparent leader when followers see that the leader's behavior will influence their physical well- being or survival" (2007).

Why isn't this just a variation of crisis leadership as we examined under Klann's definitions? Didn't Klann say that in the worst instance, a Level 3 crisis, "the loss of property is significant, the likelihood of injury is high, and even fatalities are possible? In this 'do or die' situation for an organization's leaders, public embarrassment is a given. Losses are moderate to significant, and the need to act is even more immediate."

He did say that earlier, but here's the difference. Crisis leadership and even loss of life or property damage often happens in the abstract. It happens to somebody else. Extremis leaders work in here and now, and up

Figure 10.7 Former head of the Behavioral Science and Leadership Department at West Point, Col. Tom Kolditz with the author.

close and personal. As Kolditz says about the 9/11 disaster when firefighters went up in the World Trade Center Towers to rescue people, "each firefighter at every level of command was in extreme risk while carrying out this lifesaving operation" (Pfeifer, in Kolditz, 2007, p. x).

To review and provide my own paraphrase, in Kolditz's definition, followers believe that their own well-being or survival is at risk. Klann's crisis leadership happens "over there." Extremis leadership happened "over here."

Like Collins, Kolditz first examined business cases, but he abandoned that research track because, as he says, crisis leaders are already desperate. Instead, Kolditz changed his focus to followers and what they had to say about their own leaders under "do or die" conditions.

Kolditz was in Iraq at the beginning of the 2003 invasion to overthrow the government of Saddam Hussein. He and his research team interviewed US soldiers as we might expect. These are young people in combat conditions who are highly trained individuals but ultimately reliant upon equally young commissioned officers and more senior noncommissioned (enlisted) officers who are directing strategy and operations. Kolditz and his research team interviewed 54 of these American GIs in order to learn what distinguishes extremis leaders.

Kolditz then did something rather surprising, at least to me. He also interviewed Iraqi prisoners of war, a total of 36 of them, to corroborate what he was discovering about American soldiers under the most extreme conditions imaginable. From my point of view, some of the most interesting conclusions about extremis leadership come from the prisoners, as we will see.

Back at West Point, Kolditz examined and ranked opinions of cadets who were part of the sport parachute team he supervised. He interviewed 36 individual cadets and a few team captains over three years. He wanted to compare what extremis team members and captains needed in leadership on the sport parachute team (see Figure 10.8), and what non-extremis members and captains thought. In the end, these differed a great deal as he discovered during the time his work was taking shape.

As I closely read Kolditz's work with the parachute team, I honestly had misgivings about whether sport parachuting constitutes extremis conditions comparable to a burning building. This still does give me some heartburn, I admit, however; what he discovered is that there is a huge disparity between what extremis followers want compared to everyday

Figure 10.8 Kolditz's sport parachuting team landing on "The Plain" at West Point during practice exercises.

conditions. To me, that is more important than splitting hairs about the exact nature of sport parachuting.

During his research, Kolditz invited a mixed group of people for interviews including mountain climbing guides, SWAT teams, FBI members, and emergency rescue personnel, all of whom had experienced a fatality at relatively close quarters. He interviewed 120 such people who were personally involved in extremis conditions.

In very general terms, what Kolditz suspected (and fellow sociologist and West Point professor Pat Sweeney verified during his own interviews in Iraq) were two immediate lessons. First, *competence under extremis conditions is more important to followers than trust*, which usually ranks highest under non-extremis conditions as a trait assigned to leaders. That is, under extremis conditions, followers move away from trust in favor of competence. The following table summarizes Sweeney's conclusions (see also Sweeney et al. 2011). *Trust* didn't even make the top 10 important attributes in combat. (See Figure 10.9).

In a second study, Kolditz examined *motivation* or *learning* under both extremis and non-extremis conditions. *Under extremis conditions, followers ranked "learning" highest* and not motivation of the leaders to get the job done. This is an exceedingly important finding because under business-office conditions, people look for leaders who motivate, and that's what researchers have found for years. In everyday situations, conditions are not such that followers sense their survival is at stake and the leader had better make the right choice. By definition, extremis conditions are motivating of their own accord; conditions are already critical. Instead, extremis followers in Kolditz's research were looking for someone who was a calming influence and who could spot things "out of place," to use Kolditz's words. (For an idea of the kind of leadership we don't want, See Figure 10.10: Leadership by Coercion).

Rank	Attribute
1.	Competent
2.	Loyal
3.	Honest/good integrity
4.	Leads by example
5.	Self-control/stress management
6.	Confident
7.	Courageous (physical and moral)
8.	Shares information
9.	Personal connection with subordinates
10.	Strong sense of duty

Figure 10.9 Kolditz's research suggests that leader competence is more important than loyalty under *extremis* conditions.

Leadership by coercion

How did Iraqi leaders influence their troops to enlist and fight in 2003?

"By cutting off our food, by destruction of our house, of the home, and they jailed me for a year knowing I didn't accept, that I didn't join the Army. And they tortured me."

"The Iraqis universally deserted with their weapons in hand to fight through death squads [set up to stop them], unmistakable evidence of coercion because the weapons also made them targets of U.S. forces, despite their willing desertion."

Figure 10.10 Leadership by coercion in the Iraq War.

Under extremis conditions, followers didn't want a cheerleader focused on parade-quality uniforms: *they were looking for somebody to scan the horizon for clever ways to save their lives.* Training and more training does not simplify things when situations are as volatile and ambiguous as combat or a high-rise fire. Even the best training can't cover every imaginable situation.

Under extremis conditions, followers don't want an excitable leader shouting into the emergency radio for help. Followers are already excited. Kolditz found that under extremis conditions, followers want quiet confidence and a leader who can adapt quickly.

Case study: "Horse sense" in a mining emergency

Aren't the parallels to safety and engineering leadership already apparent? I hope they are because over the years, they continue to jump right out at me. Let me provide an example. A good friend of mine and the best kind of non-excitable leader under *extremis* conditions is a man named George M. He was and still is a member of an award-winning mine rescue team for a large coal mine in our area and has decades of experience. In fact, George was the first test subject to use the rescue basket in the Quecreek mine disaster in 2001, which I have referenced in a preceding chapter. A few years ago, George went out on a rescue call where a mine roof fall had trapped three miners and took out the mine's ventilation system. There were no serious injuries that I recall, but I remember him saying that toxic gas was accumulating on the far side of the mine, making the need for quick rescue imminent.

I need to say that George is one of the best horsemen I know; I have plowed, mowed and worked my own horses with George many times, and he is among the best there is. Why is this important here? The best horsemen are calmest in an emergency such as a runaway. George gets calmer under pressure, and I have watched him calm first his horses, and then the people around him when things go bad—including a terrifying runaway with horses rocketing across a field with a plow bouncing six feet into the air. George saved the day because he never showed fear to the horses or to the people around him.

Back to the roof fall. George said nothing when the rescue leaders got all in a tizzy wondering what to do since the roof was still unstable and could fall again. The rescue leaders decided to send in a rescue robot with gas detection instruments and a radio. Two hours went by while the robot struggled ever forward, but the rescue team couldn't determine really where the robot was, only that a successful contact wasn't going to happen soon.

Another hour passed and everyone knew that time was running out for the trapped miners. George decided he should act and so he entered the mine on his own accord.

Understand that working near a recent roof fall is a terrifying *extremis* condition that only a few miners ever experience. Tons of rock up to five feet thick and a hundred feet long come crashing unexpectedly down, flattening heavy equipment and people in its path. The number one rule in all of mining has always been don't work under unsupported top. And that's just what George was facing as he crawled prone and inching his way over the top of rocks and debris on the way to his fellow miners, all within just about a 14–16 inch space. He had to work for an hour under unsupported top if he was going to rescue his fellow employees. He knew the mine, he read the methane detectors, and he knew the conditions. He also knew that there was no other choice but to chance the unstable roof wouldn't fall again.

Even the toughest extremis leaders can have a sense of humor. He told me later that as he crawled forward, he found the rescue gas detecting robot butting into a rock, recovering its robotic brain, and running into the same rock again and again, for hours. He told me he turned the robot around and aimed it back outside as a goof. George is cool under pressure.

And as you'd expect, George got to the miners with water, air packs, and a radio, and reassured the miners in the ensuing hours that they all waited. Now, the rescue operators could use heavy equipment because they were sure of the victims' location. All of the rescued miners plus George were standing outside in the sun in about 24 hours. George was brave, yes, but *competent and learned quickly*. I suspect nobody tried to stop George when he said he was going into the mine on his belly. Nobody could match his experience with mine conditions or determination, and his very presence was, and is, calming.

George was skilled and calm, and he exemplified Kolditz's notion that competence under extremis conditions is more important than trust of subordinates. Moreover, George was a leader who had learned to anticipate that in some grave conditions that the robot was insufficient despite a troop of mine safety scientists, and that a person—a human being— was going to have to intervene, and soon. That's why under extremis conditions, followers ranked *learning* highest and not motivation of the leaders to get the job done. George saw the obvious that the others missed.

Third among conclusions from the Kolditz interview research over the years, he talks about non-verbal cues. Kolditz points out that his parachute team and most SWAT Teams will shun tinted sunglasses despite working outdoors under extreme conditions. He says that a person's eyes, especially when things are at their worst, are the best nonverbal indicators of danger. The same way, he says, a leader who shares the risk with her subordinates is non-verbally elevated to a high status among followers just because she is there, taking part, sharing the risk.

In a disaster situation, followers will look to leaders to be on the spot; to do the very same things they ask their subordinates to do. Whether an *extremis* sergeant camps out among his men under combat conditions, or a business leader sets up and personally assists the command center in the event of his office building on fire, there is a nonverbal gratification among followers that this guy is in charge. No special *FBI* or *FEMA*-printed t-shirts or ball caps are needed to identify the actual leader. In fact, I am sure an authentic *extremis* leader would find the lettered and embroidered t-shirts and ball caps appalling.

I can tell you that George M would find them appalling.

Fourth, *extremis leaders share the risk with their subordinates.* They don't lead from the rear; they lead from the front. Remember the BP executive who told the press he had to go race his sailboat about a week into the Deepwater Horizon environmental disaster in the Gulf in 2010? He lasted about three more days before upper management decided he not only didn't share the risk, he was a public relations disaster.

Authentic extremis leaders arrive on site with their shirt sleeves rolled; they eat with the troops they lead. Our friend George shared the risk of his fellow miners by plunging in without being asked to do so.

Fifth, *the best extremis leaders share a common lifestyle.* Using one of Kolditz's best-of-all-time quotes, he says, "How you live shows your followers what you really value. Ideally, it's them." Does the leader attend family ball games in the summer once in a while? Does the leader visit the hospital for an injured employee? Does the leader hang around at the end of the shift and discuss, or at least try to understand, bow hunting strategies?

Anyone can see the boss mindlessly looking down using his cell phone at the ball games or looking at his watch while visiting the recuperating employee or even making jokes under his or her breath. These inadvertent but true and selfish actions are probably going to cause that erstwhile leader to crash and burn, maybe permanently. My advice is to share the common lifestyle so long as you can do it *genuinely*. As soon as the selfless action appears false, it's over.

Along those same lines, my friend George lived a simple, unassuming life with his wife, a school bus driver, and kids who are a machinist and a nurse, and, of course, horses. His lifestyle was unique in some ways, but mostly it is still indistinguishable from his fellow miners' lifestyles. He was and remains entirely genuine in his empathy for the trapped miners. He never had to say a word to justify it, and that's my entire point. Followers immediately recognize a genuine leader.

In extremis leaders seem to follow a particular pattern uncovered by his research, that is, displaying competence above all else, *being a quick learner, sharing the risk experienced by followers, and living a common lifestyle.* All of the foregoing traits of extremis leaders will enhance trust and loyalty, which inevitably feed back and foster the growth of the rest of the traits such as trust and loyalty.

Kolditz says somewhat sarcastically that trust and loyalty can't be grown by a company CEO paying for his top managers to go on a trip to the rock climbing school or rafting on a Class III river. The whole show is fake and hollow and employees will see through it immediately, and it does not develop leaders.

When the mine roof collapsed without warning, George took the initiative without being asked. He placed his own life at risk entering the mine and sharing the risk with the trapped miners. He was fully competent and skilled ahead of time; he knew the conditions and that he had a window to rescue the trapped miners. He knew the robot was merely wasting valuable time and in carefully evaluating the full situation, he learned that a new approach was going to be best. In actual fact, only a handful of people even know this story because George only shared it with his close friends, but I think it illustrates what an authentic *extremis* leader does, and because it happened on such a small scale with real people, I think we should ask ourselves, "can in *extremis* leaders like George be created? And can they train their subordinates to become *extremis* leaders?"

I think genuine extremis leaders can be created, but not in the usual way, and not without going through the soul-searching that accompanies any leader development. Kolditz says elsewhere in his book that authentic leaders are all about nurture (training) and very little about nature (being born with magical leadership traits). On this I fully agree; in fact, I elaborate on how to create extremis leaders when we get to Chapter 12 on a new kind of training for use by safety professionals and project engineers.

The process of leader development includes understanding personal values and motivations up and down the chain of command; *understanding organizational values*; acting toward employees and each other *in full congruence with these core values*; and never, never falsely bragging or calling attention to that congruence. There is one additional ingredient that the armed forces apply and the industry does not apply—and that is experiential training. We'll examine experiential training and its application to safety and engineering in the next chapter. Industry could truly benefit from it.

Summary of the characteristics of extremis leaders

Kolditz studied characteristics desired by followers under conditions where the very survival of the followers was in question. To his surprise, he noted that when things get really desperate, the characteristics that followers want in their leaders include changes from motivating and cheerleading or fancy titles to much more serious ones: *competence* ("can this guy get us out of this condition?"). Competence is ranked even higher than trust. Second was *learning* ("can this guy pick up clues and new information about our survival that the rest of us are going to miss?").

He also showed that extremis leaders are marked by *shared risk* (being on the front lines of combat or the building fire) and *common lifestyle* (if nothing else, an honest attempt to walk in the shoes of your employees). Together, extremis leaders use these characteristics consciously or not (Kolditz never really says one way or the other) to build *loyalty* and *trust* in a feedback loop.

section three

Applying leadership fundamentals

chapter eleven

What is "toxic leadership?"

The idea that leadership can be used destructively is a fairly new one in the research literature. We have all experienced a leader in name only who uses his or her position for personal aggrandizement or, even in the worst case, to abuse or punish subordinates. But young people can fall into this trap, too, and it can hurt their careers when they do not see it or attend to it immediately. This can happen when a new project engineer or safety professional jumps in a little too early with a little too much information. They have a degree and they have training, but they are extremely "green."

Before they know it, old-hand crafts people may turn against the new professionals, and roadblocks are built, inadvertently, of course, but the barriers are built all the same. This chapter discusses how to recognize and how to avoid the nemesis of destructive kinds of leadership, what is known in the research literature as "toxic leadership."

Jeffrey Lovelace (2012) says that, "Basically, society romanticizes the idea of leadership and its influence on the organization and its members. With minor exception, the majority of researchers who examine leaders, their behaviors and the outcomes they produce focus on the positive, while ignoring the negative and even destructive behaviors and influence of certain leaders."

Let's be honest: Our examination of leadership and leader development has truly been as Lovelace suggests. As a society, and even here in this book, we consider a leader to be the highest and purest form of action motivator in any organization. We have *not* discussed what happens when a leader's behaviors become destructive, consciously or not, and the outcomes are by and large negative. Could a leader act in the interest of other than his or her followers and organization?

Lovelace credits Reed (2004) and Williams (2005) with developing the notion of "toxic" leadership, a word connoting anything displeasing or poisonous. In fact, for our purposes here, poisonous leadership is the closest synonym to "toxic leadership"; the deleterious effects are felt on organizational members and, more broadly, on organizational culture and climate.

There are three characteristics of toxic leaders, according to Lovelace, who credits Reed for first noting these distinctions. First, *they lack genuine concern for subordinates.* We all know people ostensibly in leadership positions but their actions say "look at me" and not "how can I help?"

Here is a very crucial point Lovelace makes about toxic leaders in his discussion of his first point. Toxic leaders, he says, bully and intimidate because they see followers as "disposable resources they can use as they see fit."

I point this out here as something for safety professionals and engineers to be especially careful of early in their careers. Why? Because, sometimes, college graduates fresh out of school think and act as if they know more than they really do. They know the book and the content and a permissible exposure limit (PEL) or a threshold limit value (TLV), and sometimes they want everybody to know. They sometimes push their points a bit too hard, bullying to make sure followers or even peers know they are knowledgeable.

Advice for the fresh graduate: Don't shout out the answer in your first meeting

It's not that the fresh graduates have less concern for subordinates; in my experience, the opposite is true. New grads want to help move the safety or engineering functions along from day 1. It's just that they need to look around first and read the signs: Who are the information gatekeepers, who are the nominal leaders, and who is there for the paycheck only? Then and only then should the newbie proceed.

And of course, subordinates are not disposable resources; on the contrary, they are the very core—the building blocks of institutional skills and knowledge in an industry or company organization where it takes years to know all the rules and gain insights. These people are skilled and trained crafts people, not children. Even the fresh graduate knows that.

The newbie should ask more questions than make statements. Don't be the first one to shout a TLV or a static load calculated in your head. This behavior will be viewed as toxic even if it is well intentioned.

And because first impressions tend to last a long time, those artificial barriers may not come down for a long time.

The second point made by Lovelace about the work originally done separately by Reed and Williams is that toxic leaders "lack interpersonal skills or have destructive personalities which have an extremely negative effect on the climate of the organization." These destructive behaviors include gossiping, working around established and formal channels of authority, or as Lovelace says, those leaders actually "support in-fighting, [and they] abuse their informational power [structure] and behave aggressively."

Toxic leaders are leaders in name only when they act against the organization. They may pay lip service to supporting the norms and goals of the group but act otherwise when they take rumor as fact and spread it like fertilizer in the spring. An authentic leader would tamp down innuendo and rumor and gossip and not participate in the process. Here again, young professionals will soon enough be exposed to the rumor mill, and it is like quicksand. Once you're in, it's almost impossible to get out.

Finally, Lovelace says that toxic leaders focus on themselves and they act first in self-interest. In my own experience, you can spot a toxic leader by this characteristic alone. He shouts "see me do my important job."

What are the features of toxic or destructive leadership? Padilla, Hogan, and Kaiser (2007) note these features in the following list.

Toxic leadership can be characterized by the nominal leader's destructive nature or even his or her indifference, as evidenced in Figure 11.1. Padilla, Hogan, and Kaiser (2007) offer a brief checklist for determining whether a leader's behavior is actually good for the organization.

1. Destructive leadership is seldom absolutely or entirely destructive: There are both good and bad results in most leadership situations.
2. The process of destructive leadership involves dominance, coercion, and manipulation rather than influence, persuasion, and commitment.
3. The process of destructive leadership has a selfish orientation; it is focused more on the leader's needs than on the needs of the larger social group.

The endless cup of coffee: a sad tale

I once served with a toxic leader, an industry exec turned academic. Since the day I met him, he would wander into my office to talk about himself, his kids, his work, his sabbatical, his latest awards, his anything. He always carried an extra-large cup of coffee and I knew this painful interchange would last until the coffee ran out. The discussion was about him. In fact, *life was about him.*

Twice over the years, I tried to tell him, kindly and sympathetically, that his co-workers viewed him as a gossiper and strictly self-interested. I suggested that all he had to do was to ask two or three questions each time he visited somebody: "How's your proposal coming along? Do you need any help with it? How are your kids?" Ask a few questions: How hard would that be? Apparently, too hard.

In retirement, he became an even sadder case. Older now, and shuffling along the halls, he would still visit us and still he carried the giant cup of coffee.

Over 24 years, he never knew my kids' names and never once asked about them. That never changed.

One thing changed, though. People finally had enough—now they see him coming and close their doors when they know he is in the building. He still shuffles on to the next office, coffee in hand. As a former leader or in retirement, this little shuffling old man has ruined every relationship he ever had.

Figure 11.1 A quarter century of toxic leadership.

4. The effects of destructive leadership are outcomes that compromise the quality of life for constituents and detract from the organization's main purposes.
5. Destructive organizational outcomes are not exclusively the result of destructive leaders but are also products of susceptible followers and conducive environments (Padilla, Hogan, and Kaiser, 2007).

Fortunately, Lovelace does not end the discussion on the characteristics of toxic leadership. He takes us further into the world of toxic leadership by discussing how these work and how they become toxic and cites work performed by Padilla, Hogan, and Kaiser in 2007 who conceptualized the concept originally.

Lovelace says that toxic (destructive) leaders may still have vision and also charisma, according to Padilla, Hogan, and Kaiser, but they "have personalized their need for power." The fact that they do have some of the good and positive characteristics of leaders who are not toxic makes them more difficult to identify in the beginning.

How are toxic leaders supported in their own ecosystem? First, they have *susceptible followers*, again according to Padilla, Hogan, and Kaiser (2007). Those people are conformers, in their words, followers who have unmet needs and who tend to have low self-esteem. They also, these authors say, may have a low degree of emotional maturity. But while these followers are easy to influence by a toxic leader, they also probably hold a similar world view and may be ambitious, which makes this union of weakened follower and destructive leader all the more likely.

Similarly, toxic or destructive leaders operate in environments best conducive to themselves. These *conducive environments*, again according to Padilla's research team, are unstable in nature (war or organizational upheaval), pose single or multiple threats to followers, offer a direct threat to the cultural values, and operate in an environment lacking checks and balances (that is, nobody is really looking for a toxic leader who is essentially preying on susceptible followers).

Lovelace finishes up his 2012 discussion of toxic leadership by citing other startling research by Kusy and Holloway (2009) and Steele (2011), which suggests that toxic leaders are often able to not just exist, but they can actually thrive in an organizational culture that is itself toxic by nature (self-interested, operating on coercion and not persuasion). Steel says, "It is not the intent of any organization to [actively] develop conducive environments for toxic leaders, but ignorance or failure to do anything about toxic leaders enables their behaviors and can reinforce their behaviors in the organization."

Sadly, as Lovelace notes, "many abused followers develop into toxic leaders themselves." Let's keep this in mind.

How do safety professionals or young engineers guard against toxic leadership? Personal awareness instruments such as we discussed earlier in this book (the Myers-Briggs or True Colors) may help susceptible followers self-identify and make themselves less likely to be taken advantage of. Leaders should be aware of the organization's potentially vulnerable subordinates.

Routine interviews and anonymous surveys may help detect the presence of toxic leadership in a given organization. Getting right to my point here, Lovelace cites Reed (2004) when he says, "a toxic leader may be able to fool their supervisors when it comes to writing their individual work evaluations, *but they will not be able to pull the wool over the eyes of peers and subordinates*" [emphasis added]. So we ask peers to evaluate a leader's qualities, good and bad.

Toxic leaders act in self-interest, and they may be difficult to spot because they perform well in a general way. But toxic leaders prey on susceptible followers and they act in environments that are conducive anyway; they bring down the whole group and inflict morale problems. They take advantage of instability or egregious situations. In this manner, toxic leaders operate under the radar of the organization and poison the good and honorable efforts of other leaders, especially emerging ones. Only a vigilant organization with its own detection system fully deployed through surveys, interviews, and frank discussions with peers and subordinates will be ready to combat the effects of this destructive person whose own goals are more important to him than the group.

chapter twelve

Experiential training
It's not what we've been teaching in class

Safety and health professionals and project engineers alike will be heavily involved in training for the first few years of their careers. The material I have drawn upon for this chapter combines (1) the importance of hands-on training especially in potential crises with (2) opportunities to develop leaders during the training itself. The chapter represents a paper I wrote with two imminently qualified colleagues, Tom Rozman, health compliance director, Central Region Virginia Department of Labor and Industry in Richmond, Virginia, and Jim Dean, the tractor operator we met before and who serves West Virginia University as director of the Mining and Industrial Extension Department. That paper, "A Modified Model for Experiential Training for Safety Professionals and Project Engineers" was presented to the ASSE Professional Development Conference in Dallas, Texas, in June 2015, and is presented here in its entirety, except that I have edited out some material on the Millennial generation, which has already been presented in Chapter 1.

What is experiential training anyway and is it relevant to safety and engineering?

Short answer: yes it is important. Hands-on training, particularly training skills that may have life and death consequences, is an opportunity to make training stick, but also an opportunity to let junior leaders emerge.

Experiential training merely suggests learning through experience in which the learner plays an active role and training scenarios are as realistic as possible. A review of the classic trade apprentice training model that developed in Europe and continues today in a form with the German Dual System, even training methods in the Roman Army, indicate a long tradition of using experiential training to obtain high levels of skill competence. The primary use of this powerful vehicle is to develop the ability of individuals or groups to perform simple to complex tasks and function to a standard. However, the application of experiential training to safety or to engineering, particularly under extreme conditions, has taken on new interest, particularly among trainers responsible for training under the most extreme conditions. Much of this interest is underscored by fiscal and personnel constraint and dynamics.

It is my position that experiential training could, and should, be applied in the safety systems arena to optimize task-to-standard performance as well as leverage scarce training dollars. Let's explore some examples from the mining industry and the military, both of which frequently operate under extreme and ambiguous conditions where mistakes are costly in lives, equipment, and fiscal resources. At the end of the discussion I want to assess whether there are any additional advantages of training situations in a games-based environment, a training situation which should be of particular relevance to a tech-savvy Millennial generation.

Research and theoretical background

If we recall freshman psychology class, we'll remember that experiential training comes to us from the work of John Dewey and Jean Piaget, who suggested that the best learning is learning by doing. This is the addition of a hands-on experience to classroom learning; this is *application* of training.

A good example of the need for experiential training is learning to drive a car. We spend the better part of a semester in high school suffering through driver education in a static, overhead-slides kind of environment. In the closing weeks of driver education, everyone anticipates going out in the school's driver education car and practicing the cognitive skills under real-world situations. Nobody knows if the driver education car is going to encounter a bicyclist or an errant pedestrian crossing the roadway and talking on the phone at the same time—and that's entirely the point. Experiential training such as driver education is a practical application of classroom work. It is necessary for the classroom work to occur, but mastery would be insufficient. It is necessary for learners to experience real world "hands on" situations. That's the *experiential* part.

Carl Rogers, a professor at the University of Chicago, who wrote *Freedom to Learn*, now in its third printing over an incredible 40 years, makes an important observation and suggestion. Rogers suggests that the two main types of learning are *cognitive* (learning a roadmap by sitting at the kitchen table; vocabulary; math formulae) compared to *applied or experiential learning* (learning how to play a guitar or bending through a turn on a racing motorcycle). The experiential learner, goes the theory, has some control over learning and mastery outcomes (succeeding through practice, for example—no practice, no success) and being an active participant.

The best known researcher on experiential learning as it relates to organizational behavior in this century is surely David Kolb, professor of organizational behavior at the Weatherhead School of Management at

Case Western Reserve University in Cleveland, Ohio. Almost all of the proponents and even most of the grudging opponents of experiential learning (and there are, indeed, some dissenters) acknowledge the significance of his copious work over the last few decades.

Kolb says that learners will prefer to perceive and process information from the following:

> *Concrete experience*: active participants learning using hands-on activities; prefer *independent investigation* in learning
> *Reflective observation*: reasoning participants interested in whether the learning is *relevant* to their own life; prefers lecture
> *Abstract conceptualizing*: learners are most comfortable with *creating* theory and abstraction; lecture or tutorial styles preferred
> *Active experimentation*: learners are most comfortable *using* theory and abstraction; prefer drawing conclusions from observation

In Figure 12.1, we have added a dotted line to indicate where typical academic and field engineering training usually is applied. By this we mean that we usually ignore concrete/applied experience and use lecture, formulas, and theory to supply the necessary knowledge, skills, and attitudes. We grant that setting up labs or field experiences takes time and is sometimes difficult to grade. Even industry visits don't provide concrete experience. Almost always, academics and safety professionals ignore the primary component of experiential training: the actual *experience*. Trainers skip the experiential part because it's faster and less expensive to skip it, and there is low risk involved. But as our case studies suggest,

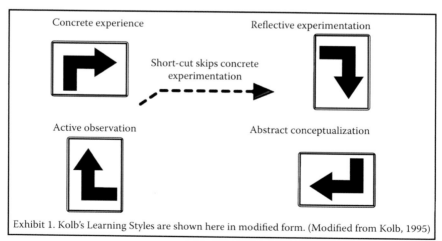

Exhibit 1. Kolb's Learning Styles are shown here in modified form. (Modified from Kolb, 1995)

Figure 12.1 Most in-class training is actually "education" because it ignores the component of concrete experience.

classroom training alone is insufficient under extreme and life-threatening conditions.

In fact, most classroom training is not training at all—it is education, yet we have over time substituted strictly classroom formats as training. It is our contention that education and training are necessary components of what organizations and professional communities do to develop individuals to agreed-upon standards in a discipline. In particular, when the situation is urgent and mistakes will be costly, experiential training becomes of great importance.

In Figure 12.1, Kolb's Learning Styles (modified) is adapted from his 1995 book, *Organizational Behavior: An Experiential Approach to Behavior in Organizations*. The figure suggests that learners prefer a particular learning style such as Concrete Experience, Reflective Observation, Abstract Conceptualization, or Active Experimentation. The typical academic setting and even later in the workforce do not include Concrete Experience. Thus, the dashed arrow we have added jumps across from the left side to the right side. The arrow connects three aspects of training, but "short circuits" actual experience.

In the typical training classroom or even the typical tailgate safety instruction, we almost exclusively feature a dry-erase board, the occasional scale model, handouts or PowerPoint® slides in an effort to achieve subject mastery or eliminate a particular performance discrepancy. In the following examples, it is easy to see that abstract conceptualization, reflective experimentation or active observation, in absence of active participation and concrete experience, would have been inadequate to save lives and mitigate property damage because a given leaner never handled a breathing device, for example, under dim and heated conditions, and became confused. To save lives and "get it right the first time" there is no substitute for experiential training.

Let's consider some case studies which make the point clear.

Case study no. 1: Comparing the Loveridge mine and the Powhattan mine fire incidents

Fighting a mine fire is a high-stakes challenge even in today's high-tech world. Mistakes are costly in terms of personnel, property, and business efficacy even when the firefighting is planned in strict accordance with basic safety codes such as 30 CFR (Mine Safety). As this case study demonstrates, a compliant mine firefighting program may not be enough to do the best job. Going beyond simple firefighting compliance will demand specialized training under simulated but high-fidelity conditions—exactly what we use in our modified experiential model explained below.

As a matter of background, in today's modern underground coal mining industry there are generally two schools of thought when looking at fire response. The first is to train each underground miner in how to use the firefighting equipment (fire extinguishers, fire hose, and "rock dust") required by the Mine Safety and Health Administration (MSHA) after notification there is a problem. Rock dust (finely ground limestone or incombustible material) may be used as a smothering agent in fire response. This school of thought is correct in that in the case of an underground mine with many remote areas and ignition sources, the best person to extinguish a fire is the individual(s) who find the fire. If these individuals are unable to extinguish the fire, mine rescue teams are the secondary response. Mine rescue team members wear a self-contained breathing apparatus (SCBA) rated for a minimum of 4 hours capacity approved by MSHA and the National Institute for Occupational Safety and Health (NIOSH) in accordance with 42 CFR Part 84, Subpart H.

The second school begins with the same thought or philosophy with the exception of fire brigades serving as a second-level response. Mine rescue teams are a longer duration response group. Fire brigade members are regular mine employees on each shift, e.g., continuous miner operator, roof bolter operator, etc., who receive specialized, highly experiential training utilizing firefighting equipment not normally found in underground or surface mining operations. The fire brigade trains under simulated conditions of heat and smoke along with ambiguous problems; they also re-train regularly. Their specialized equipment includes a short-duration SCBA along with PPE similar to those worn in the fire service, for example, Nomex hoods, fire gloves, turnout gear. The primary benefit of fire brigades is the shorter response times and higher levels of firefighting sophistication when compared to mine rescue teams.

Two very similar mine fire events clearly show the difference between using mine rescue teams or a fire brigade as a secondary fire response. In February 2003, at the Loveridge coal mine located in West Virginia, a fire ignited when contact was made with the energized trolley wire in a mine rail car hauling trash away from the mine section. The motor men operating the personnel shuttle called a mantrip, or simply "trip" shut off power and discharged fire extinguishers into the refuse car but the fire reignited as the car was moved by undertrained (but regulation-compliant) personnel to the main air course which basically fanned the flames. Perhaps worse, the same undertrained personnel moved the car away from fire hoses, and power was never cut to the immediate area. The mine ventilation rekindled the fire, the rail car was abandoned, and the mine was evacuated, all in accordance with that mine's plan. Temporary seals were established at the mine opening until the fire could be extinguished using a jet engine system. Production did not resume until August of that year. It was estimated that the total cost of the incident was six months of lost work and in excess of 90 million dollars.

In October 2009 an almost identical fire involving a mine rail car hauling trash occurred at Powhatan No. 6 Mine in Ohio with a notable and significant difference. This time, one of the motor men operating the trip was a trained fire brigade member. This time, the miner knew to pull the burning trash car into the main air course near a conveyor belt drive located in a separate entry where firefighting equipment was purposely stored. In addition, the miners had been specifically trained to immediately shut off power to the area. The fire brigade member deployed the fire hose and began to actively fight the fire by applying water. Additional fire brigade members responded immediately with advanced fire equipment deploying high expansion foam and water to extinguish the flame and cool the mine roof. The fire was extinguished in thirty minutes. No production or working time was lost; the mine was producing coal again in less than thirty minutes.

Here is the notable difference: the Powhatan mine had voluntarily purchased specialized equipment and extensively trained the individual fire brigade members through hands-on experiential training. The miners could have abandoned the burning car, but they knew they had the tools and training to extinguish the fire on the spot. They could have moved the car somewhere else, but they knew to move it close to available firefighting equipment. They knew how to use the breathing apparatus because they had practiced during training. This type of hands-on experiential training is being conducted at facilities similar to the West Virginia University facility shown in Figure 12.2.

This mine training facility is a high-fidelity experiential simulator using three mine entries and seven mine crosscuts with an overall length of 340 feet and width of 110 feet. The simulated mine contains equipment that would be present in an underground mine with mock pieces of equipment that may be configured to create various challenging and purposely ambiguous training scenarios that include smoke and heat. Even experienced miners say that this simulated mine represents as much reality as they have ever seen even in actual underground fires.

Figure 12.3 shows a simulated conveyor belt fire that fire brigade team members or mine rescue team members may experience during training. It is important to note that this simulator has extensive safety protections to prevent simulations from going out of control; that is, there are several safety features and procedures involved in this realistic type of training, including remote fuel shutoff by instructor, ignition sensing, over temperature protection, explosion proof ventilation, proper ventilation, emergency stops, and the use of proper PPE.

Figure 12.4 shows firefighting team members inside the simulated mine during an experiential training exercise. These exercises include reduced

Figure 12.2 A simulated underground mine has been constructed at West Virginia University.

Figure 12.3 A simulated mining conveyor belt fire is shown with theatrical smoke and heat.

Figure 12.4 Firefighting team members undergo experiential training exercises at WVU.

visibility, ambiguous challenges, and a changeable environment that affords many opportunities to strengthen skills needed during an emergency.

Case study no. 2: Training soldiers to maximize realism— the Abrams tank study

Toward the end of the twentieth century, the Army was confronted with a massive reorganization and integration of extensive new technologies. The purpose of the reorganization was to achieve an overmatch capability relative to potential competitors that would be two or more times more numerous if competition became active. The U.S. armored force was replacing the Army's M-60A3 tank, its principle system, with a new system, the M1 Abrams tank.

The annual tank gunnery training for an armored battalion of 52 tanks required 58 rounds per tank times the number of battalions in the active and reserve force (at the time in the range of 100 battalions or more counting armored cavalry squadrons). The new tank's gun would replace the existing main gun of 105 mm with one of 120 mm and an initial training ammunition cost increase almost double the then current cost. The amount of live training ammunition at costs per round of over \$800 required per year along with the spare parts and fuel and lubricants costs to exercise each tank on the range across the Army meant a price tag well into the billions of dollars with the existing systems. Just with the existing systems the ability to maintain life and death skill competency to minimum or better performance standards in individual tank crew members and the crews was already in question given projected budget and existing training systems and methods. This situation was exacerbated by personnel turmoil in the form of crew member replacement, especially the tank commander (TC) and gunner of the four man tank crew as soldiers were reassigned or completed their enlistments in the Army. A TC or gunner who was not up to speed, or a tank crew that did not function like a well-oiled team could mean the difference between life and death. If a training solution were not found, the United States risked a training system that would be substandard and if the nation had to use its armored, it would place thousands of Americans at deadly risk of serious injury or death.

Emerging technologies were providing the ability to create both high-fidelity and virtual-interactive simulated environments for individuals and teams to perform their actual operations as if in the real operational environment.

The primary elements of this high-fidelity experiential system were the following two parts:

The Unit Conduct of Fire Trainer (UCOFT). Three shipping containers were installed on a concrete pad in a U-Shape (later in a mobile format as shown in Figure 12.5), with one wing containing a full replica of the inside of a tank turret where the tank commander and loader could go through the entire sequence of acquiring a target, engaging it, and getting feedback on successful performance of all necessary tasks (facilitator station was installed in this wing of the UCOFT as well) exercising in a digital interactive simulation format similar to early versions of digital virtual games. The crew, as they viewed digital terrain through the replicated independent thermal sight, viewed digitized terrain and enemy vehicles that were maneuvering to engage their vehicle that they would have to engage successfully to avoid being digitally killed by that enemy. Augmented with sound effects, an environment that took on most aspects of the mission environment was created sufficient to cause individuals and collective crew to exercise to standard all necessary gunnery tasks.

Simulation Networked Trainer (SIMNET). This tank trainer (see Figure 12.6) is a facility where groupings of five tank-sized cocoon-like structures the interior of which replicated all of the crew compartments of an M1 Abrams Tank were arranged in groups of five up to a company's and battalion's worth of systems. The entire acreage of

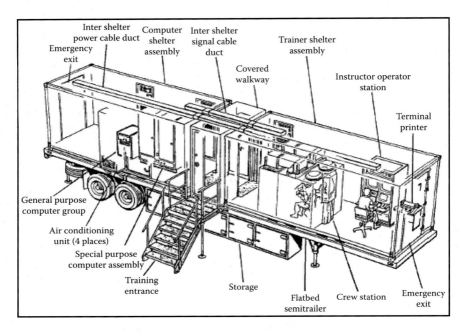

Figure 12.5 A mobile Abrams tank simulator.

Figure 12.6 An Abrams tank simulator is shown in a dedicated training environment.

Fort Knox, Kentucky, had been digitized so that the crews, platoons, and companies of a tank battalion and its headquarters could expand on what a UCOFT provided by allowing the elements of a battalion to tactically engage an enemy force in the interactive, virtual simulation environment in such way that virtually all individual and collective skills were exercised with performance feedback on success or failure, i.e., the digital enemy destroying your tank or your platoon if you failed to perform to the necessary standard.

Case study no. 3: Game-based experiential training with the Abrams tank study and with individual soldiers

Even before it was available on a commercial basis, the Army developed a tank-mounted or soldier-mounted version of what we could call laser tag, but on a much bigger scale. Even if you could say that unit was well trained in the classroom, there was, and is, no substitute for the experience of going out and fighting war games. But since it is considered bad form to fire live ammunition at your own people, there was little feedback to tell you if you were doing it right. Something more was needed.

And so MILES was invented: the Multiple Integrated Laser Engagement System (MILES). The basic MILES setup is a harness and a "halo" of laser receivers on each soldier. Each rifle has a small laser mounted on the barrel. Each time a soldier fires a blank, the laser fires a short pulse. If that pulse hits a harness, a beeper on the harness emits a loud, very annoying sound. That sound is to let you know that you are "dead." On a tank or infantry fighting vehicle, the basic application is almost identical (see Figure 12.7).

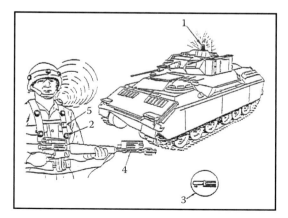

Figure 12.7 Multiple Integrated Laser Engagement System (MILES) is illustrated on a fighting vehicle and used by a soldier in training.

In full field use, the personnel of an entire battalion deploys to The National Training Center at Fort Irwin, California, drawing a battalion's worth of tanks or for mechanized infantry battalions, M2 Bradley Infantry Fighting Vehicles. In an effort to provide high fidelity without the high costs of live ammunition, the tanks are equipped with MILES. Essentially, MILES arranges a harness of sensors to the outside of each vehicle. Then it installs a device to the vehicle weapons system that replicates firing the weapon by discharging a laser pulse at a target vehicle. Each vehicle has a pylon installed at its rear. When the laser pulse hits a sensor in a location that would kill the vehicle, an electronic signal is sent to the pylon activating a "whoopee light" mounted on top and the smoke grenade also mounted on the pylon signaling that the vehicle had been "killed".

An entire enemy opposing battalion that was permanently assigned to Fort Irwin was the opponent equipped in the same fashion. All terrain being used for opposing force maneuver was digitized and when a vehicle engaged and "killed" an opposing vehicle, the pylons had a transmitter that sent the location of the engaging vehicle and the casualty vehicle to the computer for posting to the digital map, thus allowing a reconstruction of the engagement for after-action training and analysis. This system had the capability of replicating actual combat as closely as possible without firing actual rounds at an enemy.

Once the use of the MILES simulation was in full swing, and despite the effects of reduced budgets, it allowed a level of task skill training to a very high standard that produced tank crews and units that were for the most part undefeatable by opposing tank crews and units of the time. Indeed, this was demonstrated in the first and second Iraq Wars where armored forces engaged each other; U.S. forces decimated every tank battalion thrown at them with minimal losses.

The MILES training system had such redundancy that sometimes due to budget cuts, live gunnery was eliminated from a particular year's training program. The training simulations were so good that a tank battalion could be trained even without its issue of tanks.

This is not the old computer-based training where an uninterested participant merely occupies space in front of a keyboard. Instead, game-based learning and virtual training is derived from cognitive and psychological-based principles of learning in a digital age, and these concepts have proven themselves from the battlefield to the aircraft control tower.

Simulated mine firefighting in pitch dark heated environments, or soldiers and tanks "rolling on the range" and "firing real rounds" represent a full-virtual experiential simulation environment that provides:

- Immediate skill performance feedback to trainee and trainer
- Introduction of high-risk scenarios in a safe atmosphere
- Problem difficulty easily scaled up or down

- Complexity and pace adjusted for individual learners
- Scenarios that are ambiguous and challenging
- Learners who see and use the actual equipment they will use in the field

Advantages are cost effectiveness and the ability to "stop the music and review." But still, outside of certain industries, widespread adaptation has been slow, probably due to the lack of empirical or budget support, or safety professionals and project engineers who lack the particular virtual environment development skills. Also possible, organizational leadership may fail to comprehend the strategic cost benefits when compared to upfront initial investment. Yet a tech-savvy workforce increasingly composed of Millennials will probably demand this kind of training.

What about training Millennials?

The current workforce has recently entered middle management, as I mentioned in Chapter 1. But they represent a distinct population of workers and managers, as copious research suggests, including *Millennials Rising* (Howe and Strauss, 2004), *Y in the Workplace: Managing the Me-First Generation* (Lipkin and Perrymore, 2009), *Not Everyone Gets a Trophy: How to Manage Generation Y* (Tulgan, 2009), and many others. These books suggest that Millennials are altruistic and environmentally more conscious that generations past. They like to work in groups and they are technologically more literate that their parents.

But upon entering the workforce, they lack experience. And we have found the following conclusions in our own peer-reviewed and published research over the past five years on graduating OSH and engineering students:

- Lack of any real work experience prior to starting their careers can be seen as *a missed opportunity to learn about other cultures, about business, about economics, and politics.*
- Very limited outside and recreational reading can be seen as a *missed opportunity to gain economic, cultural, and historical perspective.*
- Limited out-of-state and overseas travel suggested to us that students have *This represents a missed opportunity to gain economic, cultural, and historical perspective.*
- Survey participants also knew that *learning about leadership and its practice is valuable to their own career paths including safety, engineering, social and behavioral sciences, and others.*

We know that the OSH and engineering fields, unlike most others, will someday present leaders with real life urgent and potentially crisis situations. Because the Millennial generation has precious little work experience and that the vast majority would respond to complex emergency situations by reflexively taking action by themselves with or without training, these very crisis situations might quickly become exponentially more severe.

Isn't high-fidelity, situation-constrained, technically familiar (virtual) training appropriate for all employees working under high-consequence conditions, particularly the Millennial generation? I'd have to think the answer is yes.

Developing a modified model for experiential training

There is general agreement that field practice under simulated and extreme controllable conditions, especially for those with little or no practical experience (e.g., the Millennial generation) holds promise to control losses in personnel, property, and business efficacy. For mission-critical or life-threatening events such as fire, emergency evacuation, or threats of terrorism, training conditions are purposely made "VUCA," or volatile, uncertain, complex, and ambiguous (Banks, 2010). This means that for mission-critical or life-threatening events, a truly experiential format is needed for training purposes much as we have described above.

But then training changes. Using organizational behavioral research from other sources, we add opportunities for improving unit cohesion and leader development, again with the Millennials in mind:

- Training is *introduced briefly in the usual format*: classroom and PowerPoint slides.
- Training moves quickly to a *games-based or virtual* environment.
- Training becomes truly *experiential* in the field—a seamless transition from the classroom to the simulator or field exercise (FTX).
- The foregoing activities (classroom, simulation, and FTX) must use *identical* training objectives, skill acquisition sequences and harnesses, PPE, or equipment for complete realism.
- Conditions and outcomes are purposely made *ambiguous* but not impossible.
- Immediate *coaching* is supplied by content experts.
- Written and oral *after-action reviews* are provided.
- The boss and department leaders are *active participants*.
- In the field, a *shared lifestyle* among leaders and subordinates alike establishes strong informal and formal bonds.
- *Leader development opportunities* among subordinates appear in context of experiential training when trainees meet and master experiential training challenges.

It's time to amp up experiential training and show what more can be done to make it even more effective. In this case, we can do two things:

heighten the effectiveness of the training itself, or, we can strengthen the bonds of leaders to subordinates and even offer junior leader opportunities to emerge. The best coal mine and military simulations follow this script exactly. They are called the *extremis principles*.

Kolditz (2007) suggests that in training leaders who anticipate the most extreme conditions, group leaders who participate in experiential training will both "share the risk and share the lifestyle" which provides networking and leader development opportunities in addition to modeling the correct attitudes, skills, and behaviors. And while Kolditz taught at West Point, he says trainees learn skills and leadership when bosses and supervisors are active participants. He says in his 2007 book that "Under conditions where deals may involve profits and losses of such magnitude that lives are changed forever, it makes sense that *in extremis* principles apply."

He says further, "If you are leading in any enterprise that involves risk, you need to become comfortable close to the edge of disaster and learn personally and organizationally how to handle it. It is worth putting yourself and your people at some risk to build confidence and understanding of what is required when circumstances are grave." This is the absolute essence of our modified model for experiential training: classroom with high-fidelity simulation with leader involvement.

In his October 28, 2010, Harvard Business Review Blog, Col. Bernie Banks drills home the fundamental difference between training in the military and training in business. He says:

> In industry, 90% of time is typically devoted to executing business actions, and less than 10% is allocated for increasing organizational and individual capabilities through training. The military, on the other hand, spends as much time training as it does executing—even in the midst of high stress/high risk operations. A unit in Afghanistan or Iraq will not suspend its experiential training program while involved in combat operations, because its ability to cogently and creatively address future challenges is enhanced by an enduring commitment to improving people's competence and adaptability through experiential exercises, as well as actual experiences. But the real lesson for industry leaders is not simply that training is important. What's really valuable is how the military crafts its training opportunities.

Using Banks' guidance and adding these items to our model, two other things are important to take away from the military practice of engaging in routine experiential training. First, feedback is crucial. The military practice of conducting intermediate and final *after-action reviews (AARs)*—in which all participants examine the planning, preparation, execution, and follow-up of any significant organizational initiative—fosters a learning culture.

Second, *coaching* is required to translate feedback into behavioral changes. Research has demonstrated that feedback without coaching results in little behavioral change. So, all leaders must develop their capacity to coach others. Reflection and dialog lie at the heart of development. Experiential training creates the impetus for both to occur.

Banks underscores three important things for us to learn from when conditions warrant:

The military trains incessantly, even during the lulls in combat. That's like practicing for a math test during a math test, yet they do it.

The military values training far, far more than business. If business saw "organizational improvement" or "leader development opportunities" as an outcome of training, you can bet they'd be doing it more, too. The right kind of training does affect the bottom line simply because it contains all losses. Remember the mine fire case study above? One group heavily used experiential training and suffered zero losses; the identical fire cost over $90 million for another company who trained in full regulatory compliance: that is, "by the book."

The military preaches experiential training. They don't confine training to lecture and PowerPoint because they have to be right the first time. On the contrary, soldiers and their leaders alike get down and dirty doing repetitive field training exercises.

We recognize the value of repetitive and experiential training; however, not all training justifies its added cost. Each organization must determine what skills do and do not justify the expense and time it will take. But when conditions warrant (high-angle rescue, for example), there is simply no substitute.

We see the need for a realism provided by practicing with a smoke simulator van, or a rescue boom truck. We see the need for practicing respirator fit testing under difficult and imposing conditions. And we see the need for team development and lots of coaching to provide cohesion and morale building opportunities. We also consider that the keys to success include the need for some different kind of leader development using Kolditz's model with some ideas from Banks tossed in. This is an adaptation of both Banks and Kolditz' concepts that the authors are prepared to stand on the results of their application because they work in the field.

The model presented below represents what training should look like for safety and engineering leaders—people who someday will have the lives of subordinates immediately at stake. The model derives from significant work by Dr. Winn, and is adapted, in part, from sources cited here including Kolb, Kolditz, and Banks, among others, plus the collective experience of our co-authors here. All of us know that what we teach in school is probably going to be inadequate when things go really wrong. (See Figure 12.8.)

Extremis leader characteristic[a]	Experiential training environment	Training materials and activities
Competence[a] Eventually, people recognize that this person knows what he or she is doing. They will be an *extremis* leader. Leaders model the appropriate behavior for subordinates.	Employees and leaders who will be involved in urgent/crisis preparation Examples: disaster, fire and terrorism preparation, rescue and recovery operations Starts with classroom and lab but moves quickly out to field training simulation. Field training and simulators must include physical, psychological, and environmental stressors. Realism is everything.	Lecture, PowerPoint Augmented continuously through lab and by simulation and, eventually, FTX. Repetitions are important to learn to control fear and suppress emotion during *extremis* simulations.
Rapid learning[a] Eventually, people recognize that certain potential leaders emerge as he or she learns quickly even under deadly conditions. They will be an *extremis* leaders.	Always and intentionally VUCA (volatile, uncertain, complex, and ambiguous)[b] Realism is everything.	Open-ended discussions with acknowledged content experts. Repetitions are important. High-fidelity simulation continues. Problem ambiguity challenges learners.
Shared risk[a] He or she will train with a hand-picked team, on site, especially during emergencies. They will be *extremis* leaders.	Groups of upper management plus middle and line staff join craft workers. *Everyone* erects the scaffold in training; *everyone* practices rescue techniques; *everyone* goes through the simulator.	Part classroom and part outdoors but always under expert supervision. Repetitions and variation are important.
Common lifestyle[a] He or she demands a team made up of all levels of management and she rotates the team-lead. They will be an *extremis* leader.	Shared meals and evening activities. Upper managers and even the CEO ride the bus on the field trips; all dress in the appropriate PPE. Everyone "walks the walk." Here, your résumé counts for little, quite on purpose.	Practice all components (modules) of full scale exercise. Repetitions and variations are still important and everyone discusses the simulation problem together.

Figure 12.8 An established *extremis* leader provides opportunities to create subordinates with similar characteristics under experiential training conditions.

(*Continued*)

Extremis leader characteristic[a]	Experiential training environment	Training materials and activities
Trust and loyalty[a] Teams undergo experiential training and ideally stay together for years through the efforts of *extremis* leaders.	Every foregoing activity, executed carefully, supports building trust and loyalty. Future leaders emerge because they show competence, learn rapidly, and share risk.	We purposely don't do the rope bridge and the "trust fall." They teach us nothing. Full-scale *extremis* training exercises. Repetitions and variation are important.
Feedback[b] *Extremis* leader skills are corrected with feedback from content experts.	All participants evaluate planning, preparation, execution, and follow-up. Crucial under VUCA environments.	Written feedback in the classroom and oral feedback in the field. Records kept. Repetitions and variation increase extremis leader skills.
Coaching[b] *Extremis* leader skills are fine tuned with feedback from content experts.	Coaching is verbal during training and written after training. Evaluation is done by peers and content experts.	Honest reflection and open dialog are necessary. Repetitions and variation increase extremis leader skills.

Adapted from [a]Kolditz, T.A., *In Extremis Leadership: Leading as If Your Life Depended on It,* Jossey-Bass, San Francisco, CA, 2007; and [b]Banks, C B. *How Companies Can Develop Critical Thinkers and Creative Leaders,* October 28, 2010. From Harvard Business Review Blog Network: http://blogs.hbr.org/2010/10/how-companies-can-develop-crit/.

Figure 12.8 (Continued) An established *extremis* leader provides opportunities to create subordinates with similar characteristics under experiential training conditions.

Summary of the modified model of experiential training

If we take a page from the training manual on experiential training from the mining industry and the military, we see immediately that this model expects a lot of hands-on on activity in addition to the usual classroom, at least for those employees and leaders who will be directly involved in urgent or crisis situations. Not every employee will need to be involved, but for those companies that choose this full-on preparation, such as those at the Powhattan mine, where intense, hands-on training paid off handsomely, similar rewards are quite real.

Our amalgamated model (above) for training bona fide *extremis* leaders starts with inculcating the characteristics—the Kolditz qualities—of competence, rapid learning, shared risk, common lifestyle, and trust/loyalty. Then we added two crucial components from Banks, namely, feedback and coaching, to round out our model.

Recall in the earlier Kolb discussion (Figure 12.1 above) that the standard safety and engineering emergency exercise training (the dotted arrow we added) moves across the dotted line in the figure and cheats learners out of hands-on experience. Sadly, the usual safety training in a classroom or tailgate skips the most important concrete/experiential part.

And we note that under the current system we mostly use now, classroom and even tailgate training skip the entire leader development part, too. Considering what is lost when we skip concrete experience, then further consider this: How much value can there be in computer-based training delivery when competence is required under life-and-death, *extremis*-type conditions?

We think we can do better. If we as a community of training system users are interested in training for real emergency potential (terrorism, large-scale emergency response, even a full-building fire drill), then what we are doing now in the classroom is just not enough. Cheaper yes, but do you want your high school teenager to take driver training online? Do you want your surgeon to practice heart surgery without having seen a cadaver?

For safety professionals and engineers entering the field, the quickest way to authentic *extremis*-type leadership is the use of this model or a close approximation, particularly because the young people entering the workforce have little practical experience and may be prone to making bad judgments under challenging conditions. Online training, as inexpensive as it may be, isn't going to satisfy the need in safety or engineering for the creation of *extremis*-capable leaders. Otherwise, we'll continue to train good managers who have never seen VUCA-type environments except online. This is sad and scary at the same time.

The virtual interactive and simulation capabilities that current software and hardware are capable of providing are tremendous and exciting, as indicated in mining and military examples. We are at a threshold in the training of a corporate safety inspector or engineer, and virtual interactive and simulation training environments can be developed covering every key inspection. Augmented with targeted mentoring by veteran trainer experts, we move the training of safety professional and engineer into an operational mission performance range similar to that achieved by U.S. Army tank forces in the 1980s–early 2000s. Remember that their tank training was so good and so complete and so realistic that they proved on the battlefield that they could, if budgets were cut, potentially train gunnery without using tanks, gas, and real bullets.

These training tools have demonstrated their power to raise competency levels to highly competitive and even "overmatch" levels.

Given the immense public investment in building such capabilities, especially in defense and in some larger corporate operations such as the mining, petroleum, and aviation industries on the private sector side, a currently vast public sector and private sector developmental capability exists that is capable of designing and fielding "Volkswagen" to "Mercedes" programs depending on the client's needs and budget limitations. Despite the enormous capability of these systems and their cost effectiveness after upfront costs are met, we note how little of this capability to deliver a proven modified experiential training options has been applied. This is especially true in the occupational safety and health training arena, where many organizations remain with pre-1970s era training systems, sometimes the most advanced technological distance learning capabilities being Webinar or recorded courses—essentially pedagogical formats parading as training.

What the authors here would hope begins in the community is a more aggressive exploration of the experiential training benefits—enhanced as we propose with leader development opportunities—provided by heavily experiential and virtual interactive simulation capabilities. This leverages scarce staff time, redirects precious fiscal resources, and allows smaller staffs to produce high-end competence in less time. As an occupational safety and health community oriented to saving lives, preventing injuries, and keeping organizations in business, proven training technologies and methods are available at our fingertips to affordably train our future safety and engineering professionals well beyond today's capabilities.

chapter thirteen

How authentic leaders handle the death event

I'll admit that we were wholly unprepared that summer day 40 years ago. There we were, college students working in a concrete manufacturing plant, all summer long, doing fairly routine tasks we had learned by heart. Those were dry, hot days and we were covered in dried up concrete slurry flung off rotating tubes in which the company made sewer pipes. It was the same place my friend fell backward into an upright pipe the previous summer, as if that calamity wasn't enough drama for the average college summer hire.

In my second summer, my job was working with the crane crew and moving huge metal doors to cover steam kilns where the sewer pipes cured for two days. The doors were probably 24 feet square and 10 inches thick and weighed a couple of tons each. They were massive and very unwieldy under the best conditions.

When we covered or uncovered the kilns, we hooked the crane ball to the door with a heavy wire rope, and when we were finished with the doors, we moved them to the yard some 300 feet away. Here, the doors were leaned against a welded up I-beam rack and stored until they were needed again.

My second summer there was exceptionally hot and very windy. One August day, the three of us from first shift were just glad to be going home about 3:00 in the afternoon. We had covered all of the kilns for day shift production except for one last cover still out on the rack, and we handed off this task to the incoming guys. It wouldn't take long to cover the last cover on the kiln.

Nobody saw it coming. I was already clocked out and walking to my car when I hear a loud boom. I turned around and saw a cloud of dust where the remaining steel door had hit the ground. In the gusts of August, the last steel kiln cover had been blown off its leaning position on the storage rack and toppled onto one of the second shift guys who was walking out to hook up the crane ball.

I didn't have to ask if the guy survived. It was fairly easy to see he had been killed by the weight and momentum of the steel door crashing down on him.

Nobody knew what to do right away, and it became apparent that nobody knew what to do in the intervening days and weeks either. Nobody at the company level told us anything; there was not even an announcement about the poor kid's tragic death on bulletin boards. No funeral announcement, no photo of the kid or his family, no discussion about how to prevent future incidents. Nothing that I ever saw was done to commemorate his life or prevent another instance of the same event.

I mourned the kid privately and went on back to school, and now after 40 years, I don't even remember his name. I do remember that he and his memory were just erased that day like somebody at the executive level hitting the "delete" button. From time to time, I still think about how very poorly the company handled this situation. I don't think I was scarred, but I was upset for a long, long time afterward.

Remember that we are in the business where people can die, and occasionally, people do die. We have to be ready for it to happen.

Here's another vignette. A strapping big football player here entered our safety program in the 1990s and graduated with high grades. He immediately went to work for a waste management operator and moved out to eastern Pennsylvania with his new wife. She was a gymnast, and she had blown out her knees, which made buying a single-story house an imperative, as I recall.

At the end of his first week on the new job, a bulldozer backed over a worker and killed him. My student recalls that within an hour, a news helicopter tried to land on the property, and reporters were climbing the perimeter fence to get tonight's big story. Nobody "from corporate" was there to help him. The company froze in its tracks. He was disgusted at the company's response and quit within a couple of weeks. Nothing had changed from my own college days, it seemed.

Are young professionals or staff members at personal risk today?

Now, more than ever before, safety professionals and engineers should think about planning for their organization having a fatal injury one day, and for at least two important reasons. More than in the past, they will be going or sending their staffs out on the nation's public transportation system. These will be highways, primarily, but young professionals will also be heavily using airlines, including private jets once in a while, and also the country's rail systems. The more employee-hours on the road, in any mode, the more likely is a transportation-related tragic incident just based on exposure (hours on the road) if nothing else.

When I started as a professor, most of our graduates were midcareer managers from local industry, but the student demographic has changed dramatically. Now, it is not rare at all for young professionals to be on the road three days a week, particularly in the insurance industry and, sometimes, in the hospitality business.

I worry that my own son, a young engineer, is on the road more than half of the time in his first real job. You'll remember I said earlier that the second day on his new job, somebody tossed him a set of car keys and a company credit card saying, "Your rental is outside for tomorrow's trip out-of-state."

Beyond surface transportation risks, safety professionals and engineers will be traveling overseas much more frequently and often into unstable areas where terrorism is a problem. Whether by surface transportation, air, or rail overseas, Americans are high-profile targets. A fatality through an industrial incident is no more tragic than the victim of a terroristic activity.

Business leaders should be ready for the death event. Young professionals and their staff members are at increased risk today, in my opinion, and besides, their world is a pretty crazy place sometimes.

Not a single text I have used here at the university in two decades has made mention of handling a fatality, yet those of us in the OSH business know that sooner or later, it can happen, particularly in the more hazardous industries like construction, the maritime industry, or transportation.

I have spoken about how unprepared my summer-job company was when the heavy door killed a coworker and how it affected the entire workforce for months—even years—because the company's leadership never took note or even bothered to tell the workers. I knew about the event only because he was on my shift.

Wouldn't it be better to be prepared for the difficulties a leader will encounter when there is a fatal injury at work?

An author I have already introduced, Tom Kolditz, addresses this challenging topic. In his chapter "Leading When Tragedy Strikes," Kolditz (2007) precedes a discussion on the need to prepare for the death event with a full set of needs, a list of which follows. *Extremis* leaders, among which I include every safety professional and most engineers managing construction projects, should pay attention to these times when strong and authentic leadership is imperative.

- *Hospitalization* of an organizational member or (even) an immediate family member, including death
- A *life-threatening*, lost-time accident occurring within the organization
- *Major theft or felony* crime committed in the organization
- *Significant threat* to the company's core mission
- *Legal action or credible exposure to legal action*, such as damages caused by an employee in the conduct of duties
- *Organizational exposure in the media*, such as a positive or negative newspaper story mentioning the organization or a visit to the organization by a journalist

This list of crucial events in a company's history may not even include a fatality at this point, but they are cause for special actions by authentic leaders. I attribute these topics about the death event to Kolditz, but I have shaped the following suggestions from my own experience, as you'll see.

Visiting the hospital. It has been my experience that simple but genuine expressions of sympathy and compassion are shown in the hospital visit. In my other career with a national motor sports organization, we occasionally experienced racing crashes, and as a matter of my company policy, the leaders always visited the victim. My wife always accompanied me because she registered everyone for these events and, consequently, she knew each racer, spouse, child, and the occasional race-dog as I did. The injured rider needed to get his crashed bike and gear back home, he needed to make arrangements to call in to work, and so forth. Our riders were glad for our organization's leaders to help them do these simple things, and those actions were begun in the hospital itself.

Don't wait and ask yourself "Should I go to the hospital?" Go and visit.

As a matter of leadership, the hospital visit is imperative for an organization's executives and department leaders, and especially so in the hazardous industries where serious injury is more likely, such as mining, timber, construction, and so forth.

As a policy of loss control, hospital visits go a long way to defend your organization against unscrupulous attorneys who are known to send blanket mailers to hospital rooms hoping for a lawsuit. Thus, hospital visits take on a new important dimension.

Demonstrate respect for the dead. As Kolditz (2007) sagely says, "you reach every member of your organization by the way you treat your dead…they are the most vulnerable members of any organization; they can't defend themselves." This means paying attention to information channels up and down the organization, from top executives to the last guy in the chain of command. Everyone in the organization will have questions and everyone in the organization deserves answers. Respect for the dead comes through a solid information channel.

Ritual and symbolism are important, and they have been important since the dawn of time, and respect flows through the ceremony, too. Offering a religious prayer is a kind of ritual that offers respect for the death; a prayer only has to be heartfelt—a simple prayer doesn't have to be denominational and it doesn't have to be lengthy. It only needs to be genuine.

Attending a mass, wake, or church service is another opportunity for participation in ritual by the organization's leadership. Authentic leaders will take part in these rituals to the extent they are comfortable, but attendance is mandatory.

The visitation and funeral are fairly well orchestrated and provide the family an opportunity to meet the deceased person's coworkers. The authentic leader will be at the visitation and also the funeral and, if asked, may help carry the coffin. The symbolism here is unmistakably about sincerity and comfort for the family.

There are more modern kinds of ritual that can also show sincerity and comfort outside the church. If the guy was a hunter, let the hunting symbols accumulate at his work place: his favorite hunting bow, a photo of his little brother, photos of the family on a hunting trip—these things often accumulate without sending a notice around. This also gives even the outsiders a chance to participate in the ritual. Let it happen.

Military rituals are highly orchestrated and they include flag and uniforms and ceremony. Are our business and industry brethren less important? Of course, they are not. So when we see flowers or crosses along the interstate, you can bet that the highway engineers are looking the other way for a few months or a year to let the ritual of death play itself out. They understand. We can learn from that.

Tell people in your organization what they need to know, and do it quickly. Good news or bad news, don't let CNN or the local reporters beat you to the punch. Get the basic details out right away to family and survivors first, and then the organizational leadership up and down the chain of command, and finally to the media if it is appropriate. Working prospectively shows respect to the family, and having a chance to offer a prepared statement to the media gives you the best shot at accuracy. "Don't be aloof," Kolditz says. "*In extremis* leaders can't afford it."

Meet the family in person, ahead of the funeral and ahead of even the visitation, if possible. Offer an escort for the family just to be there with them. An accompanying spouse (either gender) at the visitation or funeral can be a confirmation that the organization takes this passing seriously. Take my word for that one.

The leader meeting with family members may get pointed questions seeking to establish fault for the recent incident. That leader has to be prepared and allow as much detail as is known, but establishing fault is not why he or she is there. Rather, the leader must assure the family that a thorough investigation is ongoing and that he or she will share details as soon as these are known. The time is for compassion, not fault finding, and the leader has to turn the discussion to the former.

Take the lead by being humble. A funeral or visitation is not the time for anybody, much less a high-visibility manager, to glance at his or her watch as if he or she needs to catch the next plane. Shame on that behavior. There should be no noticeable difference at a visitation or funeral between the leader and the janitor because "the leader has to be small so the focus of the activity can make the decedent or hospitalized person big," says Kolditz. He gets it.

Giving pretty much anybody in the organization time off to attend the visitation or funeral for a coworker is good faith measure that the organization is taking the passing seriously. Naturally, not everyone will want to attend, but the offer is a genuine measure of sincerity.

Kolditz (2007) tells a good story about a funeral he attended for a group of soldiers who didn't survive an air crash in 1985. The plane carried 285 soldiers who were coming back for the Christmas holiday in the United States after a six-month peacekeeping stint overseas. The plane crashed, with no survivors, and immediately, with the high number of casualties for the high-profile Army unit, all of the Pentagon's top brass got word of the disaster.

Back home, the Army went about the grim duty of getting remains indentified and back home to waiting families. One Midwest funeral for an enlisted soldier was attended by a ranking general because of the high public profile of the crash, but everything went wrong. The American flag was folded incorrectly and sent back up the line of attending soldiers for corrections. Then a brass gun casing that is always placed symbolically in the flag got loose. It crashed noisily to the ground and fell into the grave below. Afterward, the lieutenant in charge of the funeral detail braced for the worst: He anticipated what Kolditz called a "low-yield nuclear explosion" from the attending general, but it didn't happen. Instead, the general complimented the lieutenant, emphasizing the sincerity and attention paid to the soldier's family above the correctness of the detail. The general said, "When soldiers honor soldiers, there is no such thing as a bad funeral."

In my own experience, a similar potential disaster was averted. We once had a close family member die unexpectedly out-of-state. The body was identified and shipped back home for burial. At the family cemetery, all of the funeral arrangements were made and visitation was completed, but then things went badly. Upon arriving in early morning at the cemetery on the day of the interment, the father of the deceased son noticed immediately that the wrong grave had been opened. The correct grave site was almost a hundred feet away and it was untouched—what could be more incorrect than burying a family member in the wrong grave?

But the dad did not alert the gathering crowd of the impropriety, and rather than embarrassing or chastising the funeral director for his obvious error, the father let it all slide in respect to his deceased son, in respect for the family, in respect for the process. The error could be corrected later when family and friends had all gone home. And so it was for my brother's funeral in 2005. Yes, I was pretty proud of my old man for that.

Emphasize the visual message. When a real disaster strikes and fatalities are imminent, it means the top executives need to be out front and visible as symbols of a caring organization. They are there to deliver the message to reporters who will inevitably show up in what seems like mere seconds. The message should have two parts: first, the immediate facts as they are known, and then a sincere offer of condolence. Written or orally delivered, the organization's message must have these components.

The visual message is that the company's executives are still in control and they are calling for calm in the very face of disaster. If they are sharing the common lifestyle as Kolditz recommends, they will remain at the plant, sleeping in their cars if necessary until the last remains are removed. They will probably be dressed in casual clothes rather than anything more formal. Remember the message here is entirely visual.

In West Virginia, I recall our popular former governor and now U.S. senator, Joe Manchin, on site at the Upper Big Branch mine disaster and later meeting with families of the 28 deceased miners, dressed in rolled-up sleeves and, at times, a flannel shirt. Families, parents, and spouses were angry and wanted answers. Governor Manchin, who lost an uncle in the first mining disaster covered by live television reporting at Farmington, West Virginia, in 1968, calmly let the families talk for hours and he did not interrupt them. It was a huge visual display of respect, and I could tell it was sincere. He didn't show emotion, but only real and honest empathy, and maybe that's why this guy remains immensely popular in his home state.

The likelihood is increasing for our young safety professionals and engineers, particularly those involved in the more hazardous industries where incidence rates are already high, to be closely involved in an organization's death event. They are travelling more; they are using public transportation more; and they are being employed in hazardous locations across the globe. The organization's leaders must be proactive during not only a death but also a high-profile legal or compliance action or even a significant criminal activity. There is a lot at stake, and authentic leaders need to be out front. The leader needs to act proactively to contain rumor yet still allow humble and heartfelt displays of affection to occur.

Hospital visitations and talks with grieving families and coworkers are signs of comfort and caring. Executives and department managers should meet with members before the funeral. Any hospital visit that is respectful and calm is probably good.

Respect for the dead includes allowing some formal and informal remembrance to be carried out. Photos and small displays are a reminder to the living that the memory of the dead is cherished.

Authentic leaders must get the message out proactively whenever a death or life-threatening injury occurs. The message should include facts first, then a message of genuine condolence with no emotion and no finger pointing. This is an ultrasensitive time for family members.

A funeral or visitation is not the time for posturing, but only patience. It is the time for humility. A family member will remember the safety professional attending an out-of-state funeral service a lot longer than they will remember a poor performance evaluation.

The implied message and symbolism of caring are important at the death event. Remember that George Marshall, Army chief of staff, personally wrote a letter to families of every deceased soldier in WWII before the task became overwhelming after about two years. And even though the task was later taken over by secretaries, Marshall did sign every bereavement notice. Those letters were visual messages that are recalled fondly by veterans even today.

chapter fourteen

Stress and morale challenges for leaders in safety and engineering

The effects of stress are silent, debilitating, and long lasting

Freshly minted safety professionals and engineers are going to start their jobs eagerly seeking out and mitigating *physical hazards* (machine guards, floor and walking surfaces, exits, and so forth) or *health hazards* (dusts, noise, radiation, heat/cold, and others). These are standard academic fare, and for good reason. These hazards are the ones most likely to be a risk in occupational settings.

What about risks we can't see, such as stress on the job, or organizational morale? Do we ignore them because they're difficult to quantify? Are these stressors any more significant for high-risk conditions as might be experienced by safety professionals or engineers?

The NIOSH has been aware of the debilitating nature of work-related stress for over two decades. In 1999, the institute released an important document. The NIOSH defined work-related stress as "Harmful physical and emotional responses that occur when the requirements of the job do not match the capabilities, resources, or needs of the worker. Job stress can lead to poor health and even injury" (NIOSH, 1999).

Let's be clear: There are two types of work-related stress, and they are equally harmful to the individual and to the organization as a whole. First, we have the more easily identified physiological stress and, second, psychological stress, which is more difficult to spot because the warning signs are more subjective. Both types are silent, debilitating, and long lasting. Let's deal with physiological stress first.

Physiological stress has been shown for decades to not only exist but also, worse, to internally aggravate the effects of the real stressors, such as heat-related stress or noise exposures. In fact, physiological stress for blue collar workers, as well as white collar workers, "may be an etiological factor in almost all diseases" as reported by House et al. as long ago as 1979. House's work, reported in the *Journal of Health and Social Behavior,* says that even

> perceived stress is also positively associated with reported respiratory and dermatological symptoms, but only among workers who report an exposure to potentially noxious physical chemical agents. That is, stress seems to exacerbate the deleterious effects of such exposure. The results suggest that occupational stress may affect a wide range of workers and health outcomes.

As Figure 14.1 from the NIOSH publication, *Stress at Work* (1999), illustrates, at least three independent surveys from private industry (The Families and Work Institute), an academic institution (Yale University), and an insurance company (Northwestern National Life) say similar things about occupational stress. In short, over 25 percent of workers report stress and even extreme stress at work.

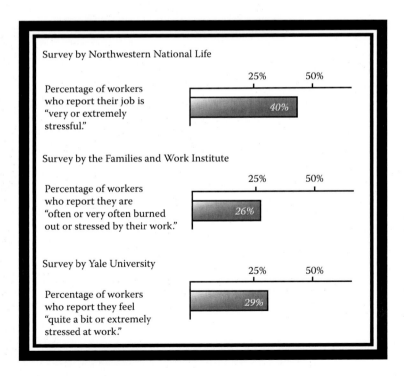

Figure 14.1 What workers say about stress on the job.

Work-related stressors trigger deep and very basic neurological and hormonal responses in humans, which are fairly consistent across individuals. The response, as we might expect, is a function of the type of stressor and the dose of the stressor. The NIOSH report goes on to say, "The nervous system is aroused and hormones are released to sharpen the senses, quicken the pulse, deepen respiration, and tense the muscles. This response (sometimes called the fight or flight response) is important because it helps us defend against threatening situations. The response is pre-programmed biologically."

Mitigating stress can be done actively by eliminating the stressor, of course, but what if that can't be done? The options then include removing the individual from the source of the stress or reducing the amount of time the individual is exposed, in much the same way we consider the idea of "time-weighted averages," which all industrial hygiene students are happy to calculate. Stress mitigation can also resolve itself passively by adding soothing sounds or colors to the work environment.

The matter of concern here is that when stress is not actively managed, the hormonal, physiological, and neurological responses continue in the body unnoticed and chronic effects begin to accumulate. The following physiological responses are extracted from the same NIOSH publication from 1999:

Cardiovascular disease: Many studies suggest that psychologically demanding jobs that allow employees little control over the work process increase the risk of cardiovascular disease.

Musculoskeletal disorders: On the basis of research by NIOSH and many other organizations, it is widely believed that job stress increases the risk for development of back and upper-extremity musculoskeletal disorders.

Workplace injury: Although more study is needed, there is a growing concern that stressful working conditions interfere with safe work practices and set the stage for injuries at work.

Suicide, cancer, ulcers, and impaired immune function: Some studies suggest a relationship between stressful working conditions and these health problems. However, more research is needed before firm conclusions can be drawn.

Workers in higher-risk public services can be affected by work leading to depression because they are exposed to more negative events (gang violence or traffic crashes, for example). A cross-sectional study of over a hundred police officers in Buffalo, New York, was initiated by NIOSH in 2007 to study the effects of occupational stress. The study comes from a long and productive line of research begun in the 1980s by John Violanti, PhD, a former New York state trooper and now professor of social and preventative medicine at the University of Buffalo's School of Public Health and Health Professions.

Studying the effects of stress on the Buffalo Police Force

In Violanti's pilot results, reported in the *International Journal of Emergency Mental Health* (2007), the investigators at NIOSH concluded that "while no association between traumatic police events and depression was observed, exposure to multiple [stress in] in life events is significantly associated with depression scores." We are left to conclude from the pilot study that having a difficult job to begin with, compounded by stressful life events, may result in symptoms of depression, and this is true regardless of ethnicity, age, and even gender of the Buffalo police officers.

The original pilot study grew significantly as it uncovered more details about whether being employed in high-risk, high-stress environments contributes to physiological and psychological stress. Uncommon in a lot of research of this type, the Buffalo study, known widely as BCOPS, randomly selected subjects in the study; accounted for age, ethnicity, and gender differences; and also compared the test group of police officers with a matched group in the population at large. This represents the very best kind of attempt at valid and reliable experimental research.

Originally, they were looking for associations between work stress and cardiovascular disease, but the study mutated, and after almost a decade, the BCOPS study researchers have information about a cluster of symptoms they call the "metabolic syndrome" among the 450-plus officers, greatly expanded from the original 100 officers in the pilot study. Having this metabolic syndrome is known to be associated with increased risk of cardiovascular disease and even diabetes.

A University of Buffalo news release about the BCOPS study says:

- "More than 25 percent of the officers had metabolic syndrome versus 18.7 percent of the general employed population
- Female and male officers experiencing the highest level of self-reported stress were four to six times more likely to have poor sleep quality, respectively
- Organizational stress and lack of support were associated with the metabolic syndrome in female but not male officers
- Overall, an elevated risk of Hodgkin's lymphoma was observed relative to the general population
- The risk of brain cancer, although slightly elevated relative to the general population, was significantly increased with 30 years or more of police service." (University of Buffalo news release, July 9, 2012)

The news release from the University of Buffalo went on to discuss the results to date with the principal investigator, Dr. Violanti, and implicitly suggested some extremely important conclusions for incoming safety professionals and engineers who will be employed in some high-risk, high-stress occupations.

"Police recruits need to receive inoculation training against stress," says Violanti. "If I tell you that the first time you see a dead body or an abused child that it is normal to have feelings of stress, you will be better able to deal with them; exposure to this type of training inoculates you so that when it does happen, you will be better prepared. At the same time, middle and upper management in police departments need to be trained in how to accept officers who ask for help and how to make sure that officers are not afraid to ask for that help," he says (July 9, 2012).

Recall that my own survey research on multischool populations (2011 and 2012) shows clearly that today's youth generally, but safety and engineers in particular, are starting their careers younger and with less practical experience than ever before. We know that their internship employers say that recent graduates are technically qualified but that they don't read about world events very much. Our current safety and engineering grads need at least some of the same kind of practical training about what to expect that Violanti suggests for his Buffalo police officers. Our young professionals are going to see and live stressful events for which they need preparation, same for their subordinates.

Psychological responses to stress and managing it

NIOSH suggests that psychological stress can stem from workplace stressors, and while there is no evidence that they result in metabolic syndrome or Hodgkin's lymphoma, they are nonetheless debilitating for the individual and the organization. The responses to psychological stress may include communication problems at work and at home; breakdown of interpersonal relationships at work and, again, at home; uncertainty of work roles; and long- and short-term career stability issues. Even such things as membership in a particular work group or environmental conditions such as crowding or noise can compound psychological stress.

According to NIOSH (1999), several studies suggest that differences in rates of mental health problems (such as depression and burnout) for various occupations are due partly to differences in job stress levels. Economic and lifestyle differences between occupations may also contribute to some of these problems.

Are these stressors and corresponding psychological responses worse under conditions with higher risk of injury or death? Indeed, they are. Another set of researchers has written about the unique challenges of high-risk conditions and they examine police, firefighters, and first responders in what they term *dangerous situations*. Patrick Sweeney is a professor and executive director of the Shackelford Leadership Institute at Georgia Gwinnett College. He coauthored *Leadership in Dangerous Situations: A Handbook for Armed Services, Emergency Services, and First Responders* in 2011 along with Michael Matthews and Paul Lester.

I interviewed coauthor Matthews at West Point in 2012 and again in 2013, where he is a professor of engineering psychology (Figure 14.2). He is the author of over 200 research papers and an active member of The American Psychological Association (APA). Paul Lester, the book's third author, is an assistant professor of leadership at West Point, from where he graduated in 1996.

Sweeney, Matthews, and Lester (2011) define psychological stress in dangerous contexts as

> Highly dynamic and unpredictable environments where leaders and group members must routinely engage in actions that place their physical and psychological well being at risk to accomplish the organization's objectives. In such situations, leaders and subordinates recognize that failures to perform their duties and accomplish the organization's objectives have the potential for catastrophic consequences

Figure 14.2 Being a professor isn't just grading papers. Here is West Point's Mike Matthews, a former law enforcement officer and research scientist on stress and morale, at a base camp on his way to the summit of Mt. Everest.

> not only for their organization but for the people it serves. The group members perceive experience or expect a threat to their well being while executing their duties. (p. 4)

We can see that this definition is almost identical to Banks' definition of the VUCA environment, where conditions are "volatile, uncertain, complex, and ambiguous," and we can also see that it is no different from Violanti's conditions experienced by the Buffalo police. Certainly, safety professionals and engineers will be faced with these situations when things go wrong, and more than ever, in the higher-risk industries and occupations such as construction, mining, quarrying, or the chemical industry, they may face these early in their careers.

The Sweeney definition given previously adds psychological risk, stress mitigation, and employee morale to physical risk under conditions where followers perceive or expect a threat to their survival or well-being. The Sweeney et al. book discusses these matters, and they are important for a fully developed understanding about what new safety professionals and engineers will face under these dangerous situations sooner or later, but especially if they enter one of the established high-risk occupations, and especially true for those involved in Emergency Medical Service (EMS), rescue, fire, or security operations. It's a good book for any bookshelf that belongs to one of our entry-level young professionals.

We know that occupations, such as police work, where the risk and stress are elevated are associated with physiological and psychological responses. The responses are automatic in the human body, they are chronic in their accumulation, and they can eventually kill you. Now, we will examine psychological stress and threats to organizational morale.

What are the psychological risk factors associated with hazardous duty? An important chapter titled "Understanding and Managing Stress" by Ness et al. (2011) speaks directly to the challenge of identifying risk factors. These authors suggest that the following risk factors "increase the probability that stress exposure will turn into a serious mental health problem. Many of these risk factors [once identified] can be modified, reduced or eliminated." Ness says that the following risk factors have been associated with a stress reaction:

- Length of exposure to operational (occupational) stress
- Severity of the operational (occupational) stress experience
- History of previous traumatic event and the amount (to which) an individual personally relates to an event
- Previous mental health problems
- Alcohol abuse or dependence
- Lack of support system or unit cohesion

Unlike conventional wisdom about stress-inducing events, which we might suspect are triggered by a single high-impact incident such as the fall of the Twin Towers in New York, Ness et al. (2011) say that, "Most stress related symptom clusters correlated with operating in a dangerous context are not attributable to a single incident. Thus, a complex of stimulus conditions within the context may constitute the stressor, which is an accumulation of events or situations outside the realm of routine that create a conflict in, or a challenge or threat to, the individual." Thus, the effects of psychological or social stressors accumulate over time and will need to be mitigated over time and, if known, brought to management's attention.

A leader's peer mentoring can lead to grit and persistence in followers

Psychological and social stressors apply in nonhazardous situations, but the point I share here with Sweeney and Ness is that *the stressors may be identified and mitigated*. In a 2009 paper, my own research team at WVU presented an adaptation of a full model of social stressors that stemmed from early work by Rhodes and Jason in 1987 and 1998. Our work was originally applied to inner-city youth development (see Winn, Jones, and Bonk, 1992, and Winn et al., 1994) but has been modified and applied again (see Winn et al., 2010, and Winn, Giles, and Heafey, 2012) in identifying and offsetting difficulties rural youth have in nontypical career selection. Lindenburg, Gendorp, and Reiskin (1993), in reviewing 35 studies using the social stress model of risk identification and prevention, have said, "According to this theory, the likelihood of an individual engaging in drug abuse is a function of the stress level and extent to which it is offset by stress modifiers such as social networks, social competence, and [social] resources."

In our most recent work, the social stress model was adapted for rural Appalachian youth for three reasons: first, because of its broad acceptance and research base nationally and internationally; second, its adaptability to multiple modes of ecological stress encountered by youth; and third, its parsimony for understanding how youth adapt to their environments and make decisions. In our adaptation, the social stress model suggests that a community by itself, or schools by themselves, or family units by themselves rarely have the resources, expertise, training, or wherewithal to support sound career decision making by youth.

In impoverished and underresourced Appalachian communities (and as we cite and attribute fully in the foregoing papers), the social and psychological stressors in rural settings include poverty, outmigration, unemployment, low college completion rates, low family incomes, high dependency rates, major industries in flux, gender dependencies, the so-called "barbed-wire" theory we created, isolation by geography, and others.

Identifying the stressors is the first step to offsetting them. The second step is building a way to reduce the effects of the stressors, and that came in the form of our peer-mentoring idea.

As we've noted in other papers produced by our research team, these psychological and social stressors will surely apply in some measure in non-Appalachian regions, but we considered that they are exacerbated in extensively rural and historically impoverished Appalachian communities. If this model holds, Appalachian social networks, social competencies, and social resources directed at making technically oriented career choices will need to be moderated by a phalanx of support and prevention methods. For example, among social stressors to youth pursuing a Science, Technology, Engineering, and Mathematics (STEM) career path, poverty and low college completion rates mean that youth will have few or no role models (e.g., friends, brothers, sisters, uncles, parents) to consult with about college living, dorm life, or how to select courses, much less STEM or engineering careers. They would not know much about technical career paths because of the cluster of stressor variables, but we also thought we could offset these influences.

With outmigration and major industries in flux in Appalachia, youth may not know which industries are moving out of their area and would have few sources of community expertise on careers that are growing or offer high-paying jobs not tied to geography. Low family incomes mean that Appalachian youth have less opportunity to afford college, and with fewer adult role models and community resources, they may never learn about available scholarships and financial aid. Managing their time poorly may lead to missing class and dropping out.

In our peer-mentoring/direct support model, we make heavy use of the influence of peers. We specifically matched peers to high-school-aged youth on variables of major and interest (engineering undergraduate student with engineering-interested high school student, circumstances of family and/or geographic region, lacking role models for technical career paths back home, local unemployment, and so forth). In this way, our goal was to overcome the stressors and help matched peers who were still in high school to make informed choices about careers and about the value of a solid career option to their family, themselves, and their community.

After almost seven years of work, the peer-mentoring model worked well to offset social and psychological stressors. As we reported in 2012 to the ASEE (see Winn, Williams, and Heafey, 2013), over 35 percent of our model-based summer camp participants went into technical majors in their freshman year. Our peer-supported campers went to a wide variety of universities and colleges outside our geographical region, and they tended to persist into (university) sophomore year at close to 80 percent retention; at least we were able to measure those effects. We are confident that identification of stressors for safety and engineering occupations is possible, and offsetting them is possible, too. Peer mentoring is one very effective way to offset the stressors. Other mechanisms to reduce stress are discussed further in the next section.

Building coping strategies to reduce stress and build resilience

Ness and his coauthors (2011) suggest the following coping strategies in hazardous situations so that leaders can offset the effects of stress (excerpted verbatim unless reduced for purposes of brevity):

- *Educate:* It is important that leaders provide accurate information so team members can set up appropriate expectations and be psychologically prepared. Constant communication and updates maintain psychological preparedness, thus *mitigating the unknown as a stressor* [emphasis added].
- *Train without Interruption:* Well-learned and *practiced skills are less likely to be disrupted by stress* than those that have not been perfected (see Dyer, 2004). In sum, stress inoculation training using realistic situations better prepares those operating in dangerous contexts for potential stressful situations.
- *Maintain Unit Cohesion:* Unit cohesion is the bonding of members of an organization in such a way as to sustain their will and commitment to each other, the organization, and the mission. Cohesive, well-disciplined units are less susceptible to the influence of risk factors than those that are loose knit and lack appropriate discipline. *Cohesion [allows a unit to make] sense of a crisis* through grounding on comrades and leaders.
- *Establish a Culture of Catharsis:* An experienced leader anticipating stress from individuals who have undergone traumatic events such as the death of a fellow team member will purposely set up opportunities to purge feelings. The goal here is to *make the feelings known but not intrusive.* The feelings are real and must not be ignored. On the contrary, individuals need to be allowed to vent. Leaders find the place and time, and shield the affected individual for a while.
- *Teach Coping Strategies:* Research shows, Ness says, that people who feel that they are in control of their circumstances and their environment feel equipped to handle the stress of hazardous situations. The administrative and bureaucratic conditions within an organization can compound the experience [of helplessness]. Efforts must be made to de-stigmatize reporting [of stressful events or conditions], facilitate support, and eliminate administrative practices that make one feel controlled by the system.

- *Commitment, Control, Challenge:* Commitment is the personal sense that one has a purpose and that one's contribution to a team is meaningful. Leaders can facilitate commitment by integrating new team members into the team, by giving them a role in it, and by giving them a sense of control through freedom to act [independently] within that role.

Operating in a dangerous context is a delicate dance between what individuals control and what happens to them. People need to experience and perceive a sense of control over their destiny, even though they are in harm's way or battling to save a life.

Ness summarizes saying, "...leaders should know their people, know the crucible, and establish a culture for catharsis [and that catharsis is natural and acceptable]. They should also be aware of the two forms of [psychological] stress producing experiences: *the critical incident and the eroding effect of the dangerous context itself*" [emphasis added] (p. 55). We have shown that we can fairly well predict uncomfortable psychological and social stresses when an unprepared high school student must improvise for college; these stresses can be mitigated (see Winn, Jones, and Bonk, 1992, 1993). We have shown clearly that the peer-mentoring and direct support model is effective in reducing the effects of social stress.

Similarly, stressors in hazardous situations take just as much anticipation, and just as much preparation, to forestall the very real organizational impacts of failed missions and debilitated individuals. Coping with psychological and social stresses can be taught in college, and it can be taught in the workplace, too.

Leaders are well advised to head off the effects of stress on individuals by identifying causes and preparing defenses well ahead of time. Absenteeism, comments by fellow workers, aberrant behavior, even a noticeable change in product quality for line workers can suggest stress. And although the HR department can help with a wellness program or referrals to healthcare agencies, identifying it isn't the job of HR; they aren't on the frontlines.

It's the job of the authentic leader.

Fostering resilience against stress

We have probably all heard at least something about posttraumatic stress syndrome (PTSD), a term used today to describe the psychological state of soldiers upon returning to more usual occupations back home. PTSD is a modern term for what used to be called "shell shock" in World War 1.

Shell shock then and PTSD both describe debilitating mental conditions that result from as a soldier being shot or shot at, seeing a colleague wounded or killed, or seeing or handling corpses (Cornum, Matthews, and

Seligman, 2011). The nonmilitary equivalent would be emergency services personnel or first responders who deal with the traumatic after-effects of traffic crashes, natural disasters, or terrorist-related events. Surely, PTSD can be applied to the after-effects of industrial incidents that have resulted in loss of life or injury.

The immediate effects of PTSD may be depression, stress, alcohol abuse, and spouse and family dysfunction, and just as with any psychological or physiological stressor, different people are affected in different ways and to different degrees. But until recently, what hadn't been different was the Army's decades long response to PTSD: Treat the after-effects. Do the screening, yes, but the response was to always wait for the illness of injury to occur and only then get involved in mitigation. But as pointed out by Cornum, Matthews, and Seligman (2011), "Waiting for illness or injury to occur is not the way commanders in the U.S. Army should approach [other] high risk actions; and it is not the way we should approach high risk psychological activities." In other words, the Army decided to be proactive off the battlefield, and deal with PTSD before it happened.

As an interesting and historical side note, Cornum and her colleagues point out the crippling effects of malaria on the civilian workforce in the early 1900s as it built the Panama Canal. "Colonel William Crawford Gorgas was detailed to Panama to deal with the massive malaria infections among workers building the Panama Canal. Employing an aggressive [proactive] preventive strategy [including screened housing, draining swamps, and the use of chemicals], Gorgas reduced the incidence of malaria from 800 cases per 1000 workers to 16" (Cornum, Matthews, and Seligman, 2011). Programs such as Gorgas' are famous for demonstrating the positive effects of proactive worker protection.

To counteract the effects of PTSD and a growing number of suicides, the Army partnered in 2009 with the APA and Army medical personnel to develop a more psychologically resilient soldier. The program, called Comprehensive Soldier Fitness (CSF), also extended to family members and to the civilian and military support personnel surrounding the combat soldier (Figure 14.3). The point of the CSF program was not to just "get better" but to "start out better."

Under the CSF, posttraumatic stress was reshaped and called persistence conflict, defined as "protracted confrontation among state, non-state and individual actors who are increasingly willing to use violence to accomplish their political and ideological objectives." The program mutated further recently with the addition of "family": The CSF has become the CSF2, which stands for the Comprehensive Soldier and Family Fitness program.

Certain including Drs. Mike Matthews and Pat Sweeney, whom we have met earlier in this book, are known for "positive" or "strengths-based" approaches. The best-known advocate for positive psychology, Martin Seligman, was a key player in the development of the CSF2. In

Figure 14.3 The logotype for the Army's CSF project.

his recent book, *Flourish* (2012), he underscores his decades-long research track that happiness and positive thinking can be learned just as depressive states are learned and not organic or medical conditions. Others such as Paul Lester made sure that empirical science was the foundation for the CSF2 program; still, others including Dr. N. Park added the importance of military children and families and the utility of the CSF2 to build more resilient soldiers even during peacetime (see *The American Psychologist Special Issue*, January 2011, for a complete review of the literature supporting the development of the CSF2).

First, the CSF2 makes sure that soldiers are physically ready; soldiers undergo physical tests and medical screening to support it. Next, soldiers undergo resilience training in groups and in web-based modules and individual training depending on test scores; the program is supported widely by "master trainers," typically senior enlisted personnel who have had combat experience themselves. The number of master trainers is over 16,000 as this book is being written.

There are five dimensions of the CSF2 program, all of which suggest that the individual affected by persistence conflict may have physiological or psychological stressors that should be addressed collectively rather than individually. The five dimensions to build proactive resilience identified in the CSF2 website include the following:

1. Physical—physical activities that require aerobic fitness, endurance, strength, healthy body composition, and flexibility derived through exercise, nutrition, and training

2. Emotional—approaching life's challenges in a positive, optimistic way by demonstrating self-control, stamina, and good character with your choices and actions
3. Social—developing and maintaining trusted, valued relationships and friendships that are personally fulfilling and foster good communication, including a comfortable exchange of ideas, views, and experiences
4. Family—being part of a family unit that is safe, supportive, and loving and providing the resources needed for all members to live in a healthy and secure environment
5. Spiritual—strengthening a set of beliefs, principles, or values that sustain a person beyond family, institutional, and societal sources of strength (see http://www.army.mil/aps/09/information_papers /comprehensive_soldier_fitness_program.html)

The evidence is trickling in that the CSF2 works. In a quasi-experimental trial of 31,000 active-duty soldiers with randomized assignment to experimental and control groups, an assessment is ongoing but a full assessment lacks at the moment. As of April 2013, assessment reports suggest that the results are in the expected direction of improvement for building psychological resilience (Harms, Herian, and Vanhove, 2013). A February 2015 website (http://csf2.army.mil/faqs.html) is now up and operating, as well as a Facebook page (https://www.facebook.com/ArmyCSF2) that shows the program continuing and having positive effects on resilience.

The Army has squarely faced the need to toughen its own workforce well ahead of the exposure to traumatic events. The CSF2 program is based on solid research that resilience to stressors can be taught and can be learned. Safety and engineering leaders of tomorrow should be aware of the need to be proactive in providing proactive resilience programs to their high-risk workforces, including emergency services personnel, police and firefighting forces, SWAT team, and others. In fact, because of the increased frequency of terroristic activities, school shootings, and even natural disasters among other traumatic events occurring in the general population, safety and engineering leaders who wish to protect their work forces from the empirical effects of psychological and physiological stressors ought to consider such a program.

A more recent (March 2014) report (see http://csf2.army.mil/downloads /CSF2InfoSheet-11Mar2014.pdf) shows three effects of using experienced trainer:

• Soldiers who received resilience training taught by a Master Resilience Trainer (MRT) improved more than those soldiers who did not receive the training, particularly in the age group of 18–24-year-olds.

- The resilience training is more effective when commanders ensure that training is properly scheduled, when confident leaders are selected as trainers, and when trainers feel that commanders support them.
- Units with MRTs had [statistically] significantly lower rates of substance abuse diagnoses (drug and alcohol abuse) and diagnoses for mental health problems (anxiety, depression, and PTSD) compared with units without MRTs.

The Department of Defense (DoD) has given full confidence in the CSF2 to continue its comprehensive work on building soldier resilience. In 2012, the first group of spouses became master trainers; in 2013, a project was launched to address teenager resilience in Army families; in 2014, the topic of nutrition earned a top spot as an emphasis area; the DoD is studying long-term return on investment through at least the year 2020.

I don't think the final evaluation is in yet whether resilience training works as broadly as we'd like, but the data available now suggest that it does.

Good organizational morale is a force multiplier

In preparing for writing these chapters, Mike Matthews and I discussed what might be most important for safety professionals and engineers about to enter the workforce to understand about how psychologists view leadership. We agreed about the need to talk about what physical and psychological stresses can do to impair the human body and the human mind. Recall that the effects of any stress can be silent, long lasting, and deadly. The effects can impact an employee's ability to work through depression, family dysfunction, alcoholism, and violence.

We agreed that young people ought to have some consideration of what everyone talks about, but, anecdotally at least, nobody does much to change: *organizational morale*. Maybe if we understood it, we actually could change it. Matthews and his coauthors again make a wonderful contribution. Their book chapter on morale, written by Reed et al. (2011), comes alive right away:

> Specifically, morale has been found to be motivating, leading to perseverance and presumably success at group tasks, especially under trying circumstances (see Petersen, 2008). Morale is potent in the face of external challenges, defined by difficulties, danger, high stress, and adversity. The defining characteristic of morale is that it is a "force multiplier"—that is, high morale has a positive impact on performance, and low morale has a negative impact on performance.

In organizations, morale entails how one thinks and feels about the group's task, mission, and purpose, which greatly affect the group's motivation to perform, especially in dangerous environments.

Let's grab hold of that Reed statement again: The author says that morale itself means almost 15 things, but mostly, it is a "force multiplier": confidence, enthusiasm, optimism, capability, resilience, leadership, mutual trust, respect, loyalty, social cohesion, common purpose, devotion and sacrifice, compelling history, honor, and moral rightness ending up as an organizational force multiplier, when taken together that's a handful. But in sum, his message is simple: Reed is saying that having high morale makes things even better under difficult conditions and that low morale makes things worse.

Under conditions of support and awareness, high morale can lead to improved perseverance under trying and even potentially lethal conditions. High morale can lead to courage and resilience (although the author does not discuss effects on risk taking, a worthy topic for future researchers especially if it is shown to have empirically deleterious effects). The take-away message is that morale directly influences group success. Morale is a force multiplier.

Reed and his associates present the table shown in Figure 14.4, modified only slightly, which indicates that morale, the force multiplier, is enhanced through leader characteristics in the left column, coupled with member selection in the right column.

Factors affecting morale

Trust between leaders and members	Member selection
Respect	Mission clarity, purpose, and moral rightness
	Task cohesion
Tough, realistic training to enhance capabilities, which boost confidence (experiential training: see Chapter 11)	Sufficient material resources
Past success; emphasizing the organization's history	Positive, caring leadership
Strong social relationships based on respect and loyalty (social cohesion)	Sacrifice for the good of the group; selfless service
Honorable performance of duty	Optimism about the future
Commitment to excellence	Devotion to the cause

Figure 14.4 Characteristics of organizational morale can be identified and modified.

Remember from Chapter 5 that Schein suggested that organizational culture consists of *artifactual values,* which you see in public but have no intrinsic purpose; *espoused values,* which your organization states overtly; and *actual values,* which may not match either of these but are the values actually practiced by the organization.

Reed's team suggests that when there is this mismatch between what we say and what we do, there is discord. The discord, Reed says, is reflected in morale problems. People act as individuals instead of team members; they are careless in what they say and do; they are selfish and they act in their own personal interest first. He says:

> Basic assumptions [of the group] can be thought of as the implicit, core assumptions that guide behavior, that tell group members how to perceive, think about and feel about things. Basic assumptions tend to be non-debatable, and hence are extremely difficult to change. To learn something new in this realm requires reexamination and reconstructing existing paradigms. *The role of leadership is especially critical to a successful reexamination and reconfiguring of basic assumptions,* [emphasis added] and therefore, to the overall morale of the unit. (2011)

We see that organizations promote low morale when their overt actions (their values in practice) do not match what they say they do. If the organization's leaders say they do have a no-tolerance policy for drugs and alcohol but they allow a midlevel manager "just one more chance," then the message downstream is eventually low morale among workers.

Imagine an office manager whose institution requires high performance and even bases all salary adjustments on strict evaluations. He then allows a worker with consistently low performance to pass annual evaluations—and even gives him a nonmerit raise—because, after all, "he is such a nice guy." That manager's decision to accept mediocrity affects only one person directly, but it affects a dozen indirectly. The message is that poor performance is OK for him, but not for you, and oh, yes, he is getting welfare instead of a merit raise. The rippling effects of inconsistency between stated and actual values can be stultifying for institutional morale, and the closer to the incongruent action, the worse the effect.

Reed offers a fairly upbeat assessment of how leaders can build morale. In Chapter 11 of Sweeney, Reed offers Table 11.5, "Leader Behaviors for Building and Sustaining Morale in a Dangerous Environment" (2011). I have made some notations to adapt its use to safety and engineering in Figure 14.5.

Leader behaviors and their relationship to safety and engineering

Reed's leader behaviors	Notations for safety and engineering
Take charge Project a sense of control Give direction Ensure continuous learning	Leader *competence* seems to the most important trait in a crisis. That comes from tough, repeated, realistic safety training to enhance capabilities, which in turn boosts confidence of everyone. Leaders will apply the "experiential training" model.
Inspire subordinates and share leadership through empowerment and participation Maintain unit integrity on missions Perform all missions in an ethical manner	Cross-training of leader's safety tasks to subordinates. Make sure leader behaviors match safety-orientation of artifactual and espoused values. Consistency matters. *Always* follow up on even simple safety-related suggestions from subordinates. Don't overlook violations just to make people happy.
Share dangers and hardships by leading from the front Communicate, explain, and live the shared values	Investigate close calls with your investigation team. Lead by simply walking around and being visible in a lost-time incident. Be first to arrive and be the last to leave. Keep an extra set of clothes and boots at the office to use when you need to "get dirty." Be first on the scene.
Remain calm Remain focused Be steady Engage in selfless service	This how a leader acts and not so much what she says or does. A positive and steady demeanor is reassuring. Selfless leaders "pay forward" and give credit to others even when things go wrong.

After Reed, B. et al. in Sweeney, P.J., Matthews, M.D., Lester, P.B., *Leadership in Dangerous Situations: A Handbook for the Armed Forces, Emergency Services, and First Responders*, Naval Institute, Annapolis, MD, 2011.

Figure 14.5 What aspiring safety professionals and project engineers can do to mold themselves—and others—into future leaders.

More than some abstract concepts about culture and values, a strong leader promotes, lives, and inspires organizational morale by being consistent and being competent, by taking charge, by sharing the hazardous duty, and being first on the scene. The leader's actions during a crisis not only reflect the organization's morale but also influence morale for the next situation that comes along. Nobody is a more central player in influencing morale than a leader who "walks the walk and talks the talk."

The leader is the central ingredient in morale being a force multiplier to an organization's safety mission.

An important update about crisis and noncrisis leader development in microenvironments or depleted environments

During my interviews and data collection, I was cautioned about attempting to suggest that leader development can occur in microenvironments or "depleted" or unsupportive environments. The cautions were well intended and honest and were meant to suggest that only large organizations such as West Point or Johnson & Johnson can actually instill a full-on leader development program.

Recall also that in Chapter 8, I discussed creating a culture of leader development even when upper management is not interested in values-congruent safety behaviors. I suggested that starting with an honor code, moving through the Be, Know, Do model of leadership, encouragement of fairly unstructured storytelling among employees, encouraging personal courage, making employees accountable and measuring their performance, and finally, mandating appropriate safety behaviors first in accordance with the James–Lange theory will spin off attitudes whose rippling effects will stimulate many other behaviors, rather than the reverse.

Sure, it's best when upper management has a huge and supportive fabric—an entire ecosystem that supports leader development and values-congruent safety behaviors, but must we despair if we work at an organization where this isn't the case? My point is simply that motivated leaders can create a supportive microenvironment if they want to. My algorithm is merely a starting point for working in a depleted environment.

Contrary to despair, I welcome the opportunity for creating a microenvironment of leader development. I don't wring may hands at all. In fact, I have a different point of view, and it is meant to suggest that small organizations or depleted groups should not abandon hope for leader development simply because they are not large, stellar, and well-known corporations. While I do think that having a supporting ecosystem is most conducive to leader development, I continue to say that what fits the large and supportive climates can also work at lower levels such as small companies, an independent safety function, or even a department within a company.

Servant leadership, or selfless service, for example, challenges us to put people first, regardless of company size. A servant leader is granted moral authority, but only on the approval of coworkers and in no other way. How can the application of these simple principles be reserved just for large organizations or simply because we have dozens of experienced military leaders as professors who say so? I respectfully disagree with those blessed with an ecosystem of support, for example, a well known multinational corporation and I mean this most sincerely. I disagree that we can't do a good job to develop leaders unless we are a large unit or because we have some special experience not shared by the general public. I am confident that a truly skilled and truly motivated leader can emerge in a department way deep in an organization if his or her stated values are dead-consistent with behavior.

In my view, leader effectiveness depends more on his or her moral character and using things like experiential safety training than on the size of the organization or its earnings-to-expenses ratio.

As I work through the representative models of leadership I have selected on the basis of my research, experience, and observation, I also believe that the empirically based characteristics of Collins' truly great companies can work at high or low employee levels—why not? I'll pick just two of them that ring true with me. First, don't rest until you have hired the best and brightest, but you also have to eliminate nonproductive and unmotivated people. In Occupational Health and Safety (OHS), that means that those who can't follow safety rules can't work for you anymore. The hedge-hog theory merely says that having an uncluttered corporate vision simplifies life at all levels: We simply do more of what we do best and skip the rest. In safety, you don't have to change paradigms every time somebody thinks of or publishes something new—work with and fine tune your own culture. Identify physiological and psychological stressors and work to mitigate them. Why can't a project engineer with a 20-person staff do that just as well as Ford or the U.S. Army? It can, most emphatically, if the leader chooses so.

I can't find where these characteristics of Collins' great companies can't also apply in unsupportive environments. A motivated safety professional can use these same principles to build a small organization where character matters and where doggedness in business makes real sense.

We moved toward crisis models of leadership and examined Kolditz's *In Extremis* leader model. He talked about competence, learning, shared risk, and common lifestyle, among other things. A safety leader can be just as effective and just as competent without being in combat.

If a safety leader or young engineer shares the risk by being first on the scene when something goes wrong, or works as a team member during disaster rescues, how is this different just because he or she is in a small outfit? In my mind, it is not different, although getting from "mediocre" to "really good" will take a lot more effort because there may be no supportive climate. I can hear it now: There is nobody else doing this "shared risk" stuff—why are we?" My point is: try it. Make it work for you.

As the best possible leaders in safety and as builders of the ultimate safety culture, we have to stay the course to become really good. That takes huge and intense motivation and likely a huge investment in time.

It may be even more difficult for a young project engineer to develop safety leadership at a construction site or manufacturing plant. After all, the industrial engineer is tasked with system optimization, not safety and certainly not "safety leadership." The civil engineer is tasked with getting an environment impact statement produced this week and a prebid package for three contractors completed by next week.

Who has time to think about common lifestyle or experiential training? But just the same, with a sense of purpose, and congruent, values-driven leadership, *it can be done.*

We examined the death event and we then examined physiological and psychological sources and mediators of occupational stress. I don't think my colleagues at any other university, occupation, or geographic region would think that their students' or employees' stresses are unique. On the contrary, the observed effects look pretty much the same to me. Why wouldn't the mitigators of such stress work at a small paint manufacturing plant with 30 employees in Texas or a mom-and-pop sawmill in West Virginia hit hard by the recent recession? The stresses are the same, the mitigators are the same, and the mom-and-pop sawmill can have just as effective leaders and a leader development program as anywhere else, given the sense of purpose needed to make it happen. Am I just being naïve?

I don't think so. I refuse to be pessimistic that a small unit or an organization in a depleted or unsupportive climate cannot make a difference about conditions and events and mostly its people under circumstances where people can die. My own research consistently shows that our youth are better prepared on the technical aspects of safety management or engineering than they have even been, period. Surely, *under conditions where people can die*, these same future leaders can make a difference simply by knowing what is required and then making the first effort.

Figure 14.6 represents a simplified model of leader development, and in particular, the right-hand box discusses an abbreviated version of the model for leader development when upper management is uncooperative or unsupportive. This is a depleted environment. We have encountered and already discussed having an honor code; we have discussed a simple model for knowing and competence, that is, the Shinsecki and Hesselbein "Be, Know, Do" model, and I add experiential training as we covered earlier. Finally, you see in the model presented in the figure that a leader can be effective growing leadership in a depleted environment through storytelling and personal commitment. So I ask those who say we need a full ecosystem of organizational support, "why can't we do this?"

The material presented in the preceding chapter sets the stage for refinement of leader development and should save young professionals a lot of time because it is summarized here. Grasping the research and sharing it is the first step; that's my job here. The second step is making it happen. That's your job.

In Figure 14.6, I have added Dr. Winn's Simplified Values-Based Leadership Model for the Depleted Environment and using experiential

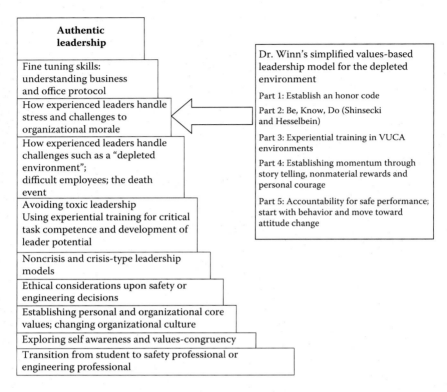

Figure 14.6 A fuller treatment of leader development with emphasis on the "depleted" environment and our modified model for experiential training.

training for critical task competence and development of leader potential and leader development in the "depleted" environment. Note how one skill is built on top of a more fundamental and lower skill until true leaders are ready to emerge and meet the challenges of tough careers in safety or engineering.

I hope you are thinking, "This just might be possible..."

chapter fifteen

Gender in safety and engineering

In the technical sciences, often known as the STEM fields for science, technology, engineering, and math, there remains a good deal of bias against women. A 2013 study titled "How Stereotypes Impact Women's Careers in Science" (Ruben, Sapienza, and Zingales, 2013) and published in the prestigious proceedings of the National Academy of Science discusses bias in hiring men or women in science careers. In their abstract, the authors summarize their findings this way:

> Without provision of information about candidates other than their appearance, men are twice more likely to be hired for a mathematical task than women. If ability is self-reported, women still are discriminated against, because employers do not fully account for men's tendency to boast about performance. Providing full information about candidates' past performance reduces discrimination but does not eliminate it.

Oddly, the bias does not reside with only males. The study also says:

> ...both male and female subjects are twice more likely to hire a man than a woman. Employers biased against women are less likely to take into account the fact that men, on average, boast more than women about their future performance, leading to suboptimal hiring choices that remain biased in favor of men. When objective information about past performance is available, it attenuates but does not eliminate the sex bias in hiring.

These findings are not news to most people, but the effects of gender bias remain, and it may even hide under well-meaning people's outward behavior.

Julian Barling, whom we have met before when we discussed transformational leadership, examines the research associated with gender roles in leadership. He cites a study of 30 years of research in his new book,

The Science of Leadership (2014), in which Alice Eagly and Steven Karau found that while men were more likely to emerge as leaders, now comes a very important lesson for my audience. They also found that female leaders were more likely to emerge "as leaders where situations where social interaction among participants increased over time." Barling explains that "one explanation for these findings is that as participants gain additional information over time, decisions concerning whom to choose as a leader are *less dependent on gender stereotypes and more likely to be guided by objective evidence* [emphasis added]."

I don't think anybody would deny that the safety profession has been a male-dominated profession for many years, but I think most seasoned professionals today with 30 years' experience under their belts would agree that there are many, many more females emerging as leaders. For a long time after WWII, safety positions were nearly exclusive to an industry's foremen and department supervisors, who, while not formally trained, understood the need for safe work practices and who were typically reliable shop-floor employees of 10 or 20 years' experience. As the safety field grew more technically demanding over the last three or so decades, and as compliance requirements demanded a broad understanding of new subfields such as industrial hygiene and ergonomics, the profession saw college graduates entering the field in larger numbers. Females slowly began replacing the retiring males in safety, and that trend continues.

My graduate classes are no different: In the late 1980s, I saw very few women enrolled, but somewhere around 1998–2000, the proportion grew to where it is about 35 percent today.

Let's apply the research now and ask why this is happening. If safety is, as I have long suggested, a field requiring a huge number of interactions and strong coordination among people and departments, and if Eagly and Karau (2002) are right that female leaders were more likely to emerge as leaders in situations where social interaction among participants increased over time, then we should be able to predict even more female safety professionals in the future just as they suggest. Eagly and Karau (2002) say that "women were more likely to emerge as leaders when the tasks were high in social complexity and where groups were large, likely because social complexity increases as groups become larger" (p. 205).

In a fresh perspective of women's issues in the workforce from the CEO at Facebook, Sheryl Sandberg has presented data on why she thinks the women's movement has stalled out, that there are still huge hurdles, but that some of the wounds are self-inflicted. "Women intentionally drop out," she says in *Lean In: Women, Work and the Will to Lead* (2013).

Sandberg says somewhat tongue-in-check that "Legendary investor Warren Buffet has stated generously that one of the reasons for his great success was that he was competing with only half of the population" (p. 7).

Sandberg sees that it isn't just the barriers of wage inequity put up by the men that have dominated these occupations. Yes, we are still paying women in 2014 only about 78 cents on the dollar that a man makes for the same work in the same job.

But she says, "Women are hindered by barriers that exist within ourselves. We hold ourselves back in ways both big and small, by lacking self-confidence, by not raising our hands, by pulling back when we should be leaning in. My argument is that getting rid of these internal barriers is critical to gaining power" (p. 8).

"Women are not making it to the top in any profession anywhere in the world," she says about her book's central tenets in a 2010 *Ted Talks* video presentation watched by over 4 million viewers. "Out of 190 heads of state in the world there are 9 women. Women make up 15 per cent of corporate top spots and that number is not moving. "

"Even among the non-profits which are usually thought of as women-oriented, only 20 percent are headed by women. Success and likeability are positively correlated for men and negatively correlated for women," she says.

She says in her video presentation that even she caught herself answering questions by men after she said that she had taken the last question. The women put down their hands, but many of the men did not. That event set her to thinking how ingrained these biases are—even she had them.

So what does Sandberg mean by "lean in?" She provides about six rules that women should follow to help secure their future. One of them, "sit at the table," means for women to not sit at the side of the room where they typically do in deference to the men at the same meeting. This puts them out of eye range for questions and makes them seem more distant psychologically because they are more distant physically.

Another tip Sandberg suggests is, "don't leave before you leave." This means, and her data show, that women will often telegraph their plans to change jobs long before they need to, and far longer in time before the men indicate a willingness to move on. Instead of "leaning back" and signaling their desire to move on, women should play their cards "close to the vest," she says.

Maybe it's too severe to say that based on the Sandberg book, "women are their own worst enemy," but it's probably true that they could stand to take a fresh look at their own motivations and how they act in social and professional situations. I suggest the Sandberg book to get started down this path—and I suggest it for both women and men.

What about the numbers? How many women are out there in the respective fields of engineering and safety?

In safety and engineering, the discrepancies in college enrollments are the subject of many a state initiative to increase the ranks of women in STEM, and surely, safety plus civil or mechanical engineering are

STEM fields. The National Science Foundation's (2012) "Science and Engineering Indicators" says, "the number of science and engineering master's degrees earned by both men and women rose between 2000 and 2009, but the number for women grew slightly faster. In 2000, women earned 43% of all S&E master's degrees; by 2009, they earned 45%."

In safety, the proportions of women graduates and those with the CSP designation are similar to those in engineering.

Brian Yoder reported these figures in 2011 for the American Society for Engineering Education (ASEE), clarifying the proportion of women to men. Surprisingly, biomedical engineering has only been a popular field in engineering for less than a decade, yet it is second highest to environmental engineering in 2012, the date of this report (Figure 15.1).

Yoder comments for ASEE on these data saying that females accounted for 18.1 percent of U.S. bachelor's degrees in engineering in 2012, growing only slightly to 18.9 percent in 2013, a number that has essentially remained unchanged for a decade or so. For masters' degrees in engineering, females accounted for 22.6 percent of all such degrees and slightly more at 22.9 percent for doctoral engineering degrees.

Women among engineering faculty account for 13.8 percent of all tenured and untenured faculty in the United States, according to the same Yoder report. This number is increased from 9.2 percent in 2002.

The number of science and engineering degrees peaked in 2009 at almost half a million, according to "Science and Engineering Indicators," and has been growing steadily for about 15 years (NSF, 2012). While men do earn more engineering degrees overall, plus computer science and physics, women graduate more frequently in the following STEM fields: chemistry, biology, agricultural science, sociology, and psychology.

ASSE's Kim McDowell, director of Member and Regional Affairs, and Sue Trebswether, editor of ASEE's Professional Safety, offered these descriptors of females in the safety profession. We note immediately that the proportions are almost exactly the same in the engineering fields (not college graduates), just under 20 percent (Figure 15.2).

The following descriptors suggest that females continue to enter the safety profession and have become more racially diverse in recent years:

- ASSE's female members are younger than male members.
  - The average age of an ASSE member is 47.6.
  - The average age of a female member is 43.
- Most females are credentialed.
  - 18% of female members have CSP.
  - 26% of female members are professional members.

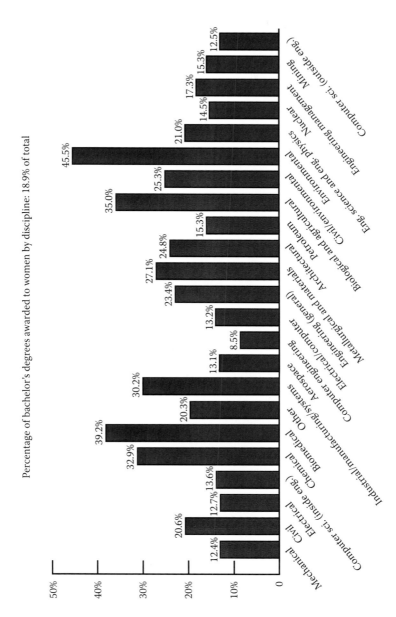

Percentage of bachelor's degrees awarded to women by discipline: 18.9% of total

Figure 15.1 Engineering degrees go to only about one-fifth of graduates as late as 2013.

ASSE historical growth (2010–2014)

	Total membership	Female members	Male members	Unreported	% of female members
2014	36,003	6,391	26,898	2,714	18%
2013	34,967	5,976	26,439	2,552	17%
2012	33,957	5,789	25,724	2,444	17%
2011	33,237	5,452	25,363	2,422	16%
2010	31,747	5,106	24,419	2,222	16%

Figure 15.2 ASSE reports about the same proportion, one-fifth, for female safety professionals.

- Most females have a technical education.
    - 48% of female members' highest education is a bachelor's degree.
    - 28% of female members' highest education is a master's degree.
    - Growing diversity in female membership: in 2013, 77% of female members were Caucasian, dropping by 2% in 12 months.

Source: American Society of Safety Engineers, 2014

Gender observations from the frontlines

Especially among older safety and engineering pros, there is a perception that females entering the safety professional will find hostility on the shop floor, in meetings with project owners or general contractors, or in the gas field. While there is apparent truth to that statement, there is also truth to the idea that women in the engineering and safety fields can, and do, alter the safety and engineering cultures for the better. They have to adapt, yes, but they also thrive.

Jennifer Worthington works for a drilling operator in central West Virginia. She comments about her position in the gas industry (Figure 15.3):

> I can tell you from experience: this is not the place or the industry to be sensitive and easily offended. I have been called every name in the book, heard words that would make sailors blush, and have even experienced attempts of men from all walks of life being flirtatious and attempting to walk away with a date. In order to survive, I had to learn to be firm and walk a very fine line of earning respect from these men but not allowing myself to be disrespected at the same time.

Figure 15.3 Jenn Worthington at a drill rig where she is the ranking safety professional.

I accomplished this quickly by not taking any flack from those who showed disrespect and tactfully put them in their place, sometimes in front of their buddies if that is what the situation allowed. I have learned to walk confidently and shake hands with a purpose because your first impression will last and it has to be understood that you mean business.

I have learned that admitting you are wrong, and at times you will be, shows strength not weakness and I take great pride in knowing that I have mastered the art of making a full grown man cry without ever raising my voice or slurring any profanities; there is nothing more intimidating to a man than an emotionless woman. I will be the first to defend my people when the situation calls for it but I will also not hesitate to ensure someone is aware they are putting lives at risk and expect them to put forth the effort to correct the behavior.

I have always been blessed to find a seasoned veteran of the industry with a head full of gray hair, scars to show and stories to tell take me under their wing and teach me, show me, and gladly answer all of the questions I could muster up at any given moment. I would always make a point to find someone to fit that description and make friends with them immediately, no matter how grumpy they seemed on the surface, because it didn't take me long to figure out these were the men that lived through the industry when it was not so closely monitored, and managed to live through the mistakes they made that nearly killed them. All of them can recall instances of serious injuries and close calls when they were young and most can give you the names and tell you about the family of a fellow roughneck killed in the field.

Ava Dykes earned two degrees in chemistry and then a doctorate in biomedical sciences while raising two small children and working part-time on her own (Figure 15.4). Now Associate Service Fellow at the Toxicology and Molecular Biology Branch at NIOSH, she has experienced gender bias face-to-face. She says:

The notion of male superiority in the sciences is a long-standing prejudice that many women, perhaps subconsciously, meekly accept and learn to cope

Figure 15.4 Dr. Ava Dykes operating a high-resolution transmission electron microscope.

with. For example, I recently read a New York Times article about gender bias in the sciences (see: Chang, Sept. 25, 2012). The author reported on a Yale study that found when equally qualified candidates were presented to test subjects, men were more likely to be hired for a job and usually paid up to $4,000 more for similar positions. It seems as if gender biases follow us even decades after Gloria Steinem made all the papers.

After accepting a new position involving a technical subject area that I hadn't worked in before, a colleague made the comment that maybe I was hired [to operate to a high-resolution transmission electron microscope] simply because I was a blonde. I think his comment was meant to be flattering, but his statement had the opposite effect. The idea was insulting that my qualifications and scientific knowledge weren't the basis for my career progress thus far. The desire to prove that particular person wrong caused me to work even harder, and eventually became a springboard for my success in this field.

It seems a cliché message to young women entering technical fields, but doing your best and working hard are still the only means to accomplishment. While it's true that a pretty face can open many doors, the right ones can only be opened by showing your expertise and proving your value regardless of what stereotypes you have to overcome.

Second Lieutenant Laura Dukens is a recent West Point graduate with a mechanical engineering degree. She never made excuses for being one of only a few women in either engineering or West Point; instead, she adapted and thrived. When she was too small to get over the 12-foot barricade, she asked her teammates to grab her shoulders and pitch her over. It became their time-saving strategy for the next two years (Figure 15.5).

Laura has this advice for new female engineers:

- Don't try to be "one of the guys"; it makes you look phony.
- Set goals and do it regularly. I always heard, "shoot for the moon, and even if you miss, you'll land among the stars." OK, it's cheesy, but I believe it.

Figure 15.5 2Lt. Laura Dukens is a combat engineer and recent graduate of West Point in mechanical engineering.

- Find friends and network like crazy. True friends support you, defend you and define you.
- Engineering and the military are difficult for anyone, even men; so many people will respect what we're doing in helping people everywhere in the world. Let's let our resumes speak louder than our chromosomes.

Aly Castro owns a hot-rodded Dodge Charger and two industrial engineering degrees (Figure 15.6). She is a high-level quality control engineer at Bayer Corporation in Pittsburgh and regularly travels to the company headquarters in Germany. She has a long list of academic and career honors going back almost two decades. Currently, she is also a senior examiner for the Baldridge Performance Excellence Program, which "helps stimulate American companies to improve quality and productivity for the pride of recognition while obtaining a competitive edge through increased profits" (http://www.nist.gov/baldrige/about/improvement_act.cfm).

Figure 15.6 Aly Castro receiving the President's award at Bayer HealthCare in 2014.

Aly doesn't take any lip from people at a stoplight or at work, either, yet she is somewhat jaded about workplace gender discriminations. She says about discrimination, "It's still there, and in a Hispanic culture, it's really be out in the open." She says:

> After performing the Quality Director functions for a year and a half, when the time came, they appointed a male engineer that had no experience in operations or in management. For me, that has been the most painful experience ever.
>
> I would not say biases are increasing, but certainly they are not diminishing. Bias is still very latent—an undercurrent, so to speak. And the truth is, women with more capabilities than men, performing the same job, are often paid less. I have even seen a human resources manager interviewing a female engineer and asking if she had kids, if she had been married, and if she was planning on having a family. A human resources manager should know, as I do, that these questions are entirely unethical and in fact, they are illegal, but it has happened in front of me.

I asked Aly what advice she might have for young female engineers, and for young male professionals, for that matter. Not surprisingly, she had advice on some topics that I have heard from other graduates about three to five years out in the field:

> First, I made the mistake of not networking very early in my career thinking that performance alone was the only thing needed. Big mistake! If leadership does not know you at a personal level, you will miss a lot of opportunities. Even without the initial networking I managed to be successful, but it was way more difficult than it should have been if I had interacted with people more. Second, let your hair down and don't be too square. It will only hurt you. Here is my third bit of advice for the young women. Don't show your emotions, as difficult as it may be.

Summary of research associated with gender issues

On the research side, we reviewed Barling's work concluding that female leaders were more likely to emerge where situations with higher levels of social interaction among participants increased over time. Barling explained that leader selection in these cases is less dependent on gender stereotypes and more likely to be guided by objective evidence—job performance, for example.

From the National Academy of Science, and others, we saw where the number of females entering sciences and engineering is up slightly, as are the numbers for female faculty in undergraduate and graduate programs, but the total numbers are still rather small at about 19 percent for engineering and 22 percent among safety professionals. We do see females earning more than half of the undergraduate degrees in the United States at 57 percent, and close to half of master's and doctorate degrees at about 43 and 45 percent, respectively.

In Sheryl Sandberg's recent and immensely popular book, *Lean In*, she wryly observes that the women's movement has run out of momentum, that discrimination is still rampant, and that many of the injuries are what I called "self-inflicted." Yet far from being dismayed, Sandberg has simple advice for young professionals: Women need to know about the biases they may experience and men need to recognize the biases and help rid the workplace of them. Women don't need to be rude, but they do need to be assertive, she says, especially when they know the answers but they are simply discouraged from participating.

On the anecdotal side, I selected four strong, career women who are just as likely to dazzle you in the horse show ring or doing a burnout at the drag strip as they would accompanying an OSHA inspector or talking at the nanoparticle scale. All four women know from firsthand experience that gender biases exist in the workplace and that they are often subtle, but each has made peace with this reality and developed coping strategies. Jenn Wuchner suggested partnering with an old hand at the gas drilling site not only to learn the issues first hand but also to break down barriers among staff through trust and accurate understanding of safety issues. Ava Dykes says her own work output increased over the years to offset bias, and just so her performance would be judged equal to her male counterparts. Laura Dukens wants her work record to be the deciding factor, not whether she's a girl in a man's engineering world. Aly Castro suggests that women should network early in their career, fit the local culture, and try not to show emotions. Each of these women recognizes the existence of bias and has consciously strategized to offset it; their obvious career success suggests to me that strong women will succeed in engineering and safety careers despite the existence of gender bias, even if it takes more time than if they were male.

There are two final notes to discuss here about gender. On the first, I asked my classes over many semesters, both the men and the women, if either gender was offended by the use of terms *guy* as in "you safety guys" or "don't be that morning guy walking around with coffee." Not only were they not offended, but also many of the women in my classes said they already use the term themselves. I chalked this concern off my "list of concerns." Apparently, the use of the term *guy* isn't offensive, and while my unscientific poll clearly isn't the last word on the subject, it does also to be contextual. That is, their professor was not hostile in posing the question, and the subjects were all Millennials, who seem less concerned about it than older professionals are, at least in my experience.

Second, and in a broader context of influencing safety norms outside work, I want to thank another of the strong women in my professional life for the following two observations. Hillary Strawser-Dean is the senior safety professional at a large surface-coal company and oversees other separate mines. She and her family also run a medium-sized farm in central West Virginia. She's up about 4:30 every day to feed the stock, and she gets back to the farm about 7:00 p.m., if she's lucky. Hillary pointed out that women are in the unique position as mothers to pass along safety training to their own kids and maybe other local moms. "Stay back from power take-off shafts" or "Use safety glasses when we cut firewood."

Enter the "safety mom." Hillary says she uses the idea not only at home but also at work.

I know this is not really new, but it does seem to me that a strong but trained female influence for safety at home should be associated with higher incidence of safe behavior later when those kids grow up. Moms are huge influences when they're directly involved in their own kids' safety; those kids grow up already acclimated to a personal safety credo.

Hillary also commented on what she calls the safety mom culture at work. "When women safety professionals have fairly free reign to create programs, do training and set up safety committees, there is abundant worker and employer interaction," she said. Hillary also said that earlier in her career, she found herself establishing a nurturing culture different from the one she learned working with a male safety pro. She became a safety mom, and it has worked, even with a few drawbacks here and there. She listens more and yells less than her male predecessor does, at least that's what she tells me. Apparently, it works.

In actual fact, the research data may support the notion that men and women create different yet equally effective safety cultures, so Hillary's safety mom culture might have been predictable. Dr. Emma M. Seppala (2014) recently summarized salient research literature in her blog on *Psychology Today's* website titled "Feeling It: Emotional Expertise for Happiness and Success." She noted that researchers find activity in different parts of the brain comparing women with men when test subjects experience compassion. It suggests to me that the manifestation of safety cultures might also be different by gender.

Seppala says about these gender differences:

> Another reason women may have learned to express compassion more easily emerges from the work of Shelley Taylor, at UCLA, who found that men and women respond differently to stress. These differences may have certainly trained women to express compassion more explicitly. Taylor found that the "fight or flight" response is characteristic of men whereas women tend toward a different tendency: "tend and befriend." Women faced with a stressful situation are more likely to respond by socializing, bonding with others and seeking protection and nurturance within a community.

These tendencies may have been evolutionarily adaptive since we have evolved in communities where women's primary responsibility was raising and protecting offspring who needed protection while men traditionally engaged in hunting and warfare. New studies, however, suggest that men, too, also can respond to stress through social bonding.

It appears that Hillary discovered on her own what psychologists such as Dr. Seppala are finding in the lab, that men and women may well establish effective safety program but execute them differently. There's probably a master's thesis or dissertation topic right there staring at us.

chapter sixteen

How authentic leaders handle the issue of discipline for difficult employees

It did not take long in my experience as a new professor to see that students are really afraid of knowing they will have to discipline employees someday. Really afraid. Here are young college graduates standing toe-to-toe with a hardened old carpenter or steelworker about him or her not following the rules.

And then I remembered what "Captain Jack" taught me when I worked at the helmet factory. It doesn't show up in the textbooks I have used for class, but it works because my students have told me so.

I had not heard of *progressive discipline* until I went to work as an industrial engineering in a helmet manufacturing operation in the late 1980s. My direct boss was a retired Navy captain named Jack C, and after a decade of retirement from the Navy, he still ran a tight "ship" there in Illinois, 1500 miles from any ocean. Jack explained that progressive discipline is a handy and rational way to apply increasingly severe levels of behavior control whenever there is somebody who doesn't get it or doesn't want to get it. I also note that in the texts I have used in my time teaching introductory safety or engineering principles, I have not encountered a discussion of progressive discipline applied to either area.

The need for progressive discipline is clear. First, employees have a right to know exactly what behaviors they are doing that are not performed according to the training performance standard or Job Safety Analysis. Discipline comes after training, at such time when the employee can't or won't perform. Progressive discipline is a way to shape behavior through coaching, and it minimizes embarrassment for the employee. Finally, your human resources (HR) department will want, even require, you to go through these steps as part of the company's due process, meaning that the company owes it to each employee to make sure nobody jumped over disciplinary steps for any reason; that is, the company applied its "due process" as the law requires. And because many leaders don't like confrontation, progressive discipline gives you a structure to move through stepwise and carefully. Here is how it works.

Case study: The drill press operator

Let's say you have an employee, Lisa, who repeatedly fails to lower the Plexiglas shield to protect her face and arms from shards of a broken carbide drill bit should it shatter during aggressive drilling. The same woman runs her drilling machine faster than the speed recommended by the manufacturer and your plant industrial engineer. The employee is turning out parts with the correct specifications, but she is endangering herself and her scrap and rework rates are higher than anybody else— way past the level specified in the production standard.

This employee may be injured, and the motor and bearings in her drill station may prematurely fail. She is the same employee who doesn't understand that she is jeopardizing her own safety and running her machine right into failure. What does the safety professional or plant engineer do? They call into play the idea of progressive discipline, a way to shape the behavior of nonachievers in rational steps.

Step 1: *You counsel the employee on the shop floor* but not publicly and not to cause embarrassment. Here, early in the process, the safety leader is firm and fair but a bit casual. Talking your way toward behavior change at this step is a lot more cost-beneficial than going further with written reprimands, and besides, Lisa might have just been unaware of the issue of the shield and machine speed. While you have decided that her unawareness is unlikely, it is still possible. It's easy to correct the unsafe acts at this stage.

As the plant engineer or safety professional, you make sure that you ask Lisa if she needs the shield to be rebuilt or if the machine's speed control is working ok. If the shield or drilling machine is not working correctly, you will be applying discipline without actual cause, so you identify and fix the root cause of the problem and everyone goes back to work. But if the machine and shield are ok, the root problem is her behavior.

At this early stage, you'll want to apply a little bit of coaching without a written record. You ask her if she understands what is going on and that you hope to solve the matter here and now either by fixing the shield or speed controller or by focusing on the behavior. However, you clearly and firmly point out that at the next stage, things will get more serious all around and there will be consequences for failure to comply.

The speed controller works and the shield works just fine, too. Lisa continues to run the machine hard and not use the safety shield. She doesn't get it. *She is one of Dr. Winn's classic nonachievers.* What to do?

Step 2: *Call the employee in to your office later in the afternoon for a meeting.* If you give Lisa a couple of hours to think about the meeting, that time helps cement the idea that you are serious about changing her behavior if it was unclear for any reason before. When she comes in, check with your HR staff because you'll probably need a witness in your office—another employee who can be neutral and to make sure there is no gender bias unfair intimidation.

At step 2, you'll again discuss the matter fully but this time you'll have Lisa sign a written reprimand that will go into her permanent personnel file in the HR office. Maybe she refuses to sign, but you proceed anyway, noting "employee refused to sign" and proceed. You coach her about the need to protect herself and the need to protect company drilling machines. Tell her about the costs of scrap and rework and that her scrap rates are higher than anyone else in the department. This time, you tell her that if you observe the behaviors again, you will be suspending her for two days without pay in accord with the company HR policies. You ask her if she understands what is going on and hope that she does understand. If the behavior corrects itself, our work here is finished. But ultimately, if she does not want to change her behavior, then her work here will be finished. Those are the ultimate consequences that she must appreciate.

But still, Lisa's self-destructive behavior continues. Even after a verbal warning and a private counseling session with a formal reprimand, Lisa continues to run the machine hard and not use the safety shield. She still doesn't get it. *She is now truly a nonachiever.*

Step 3: Another meeting and, this time, suspension. At this point in the steps of progressive discipline, it is obvious to everyone that Lisa needs to be sent a stronger message. Her continued path ignores the obvious concern you have expressed toward correcting the behavior, but your concern is ignored for some reason. Here's Lisa's one last chance.

You call her in one more time and provide another written reprimand. Again, you call in a witness, again you ask for signatures, and again you ask her if there is anything you may have missed that is at the root of the problem. But this time, your coaching includes a new outcome for the future. There won't be any more warnings after today; there is only termination staring at her and she has to understand your position.

Surely, most people get it by this stage of progressive discipline. But apparently, Lisa does not.

Unfortunately, you see that Lisa's path toward self-destruction continues to play itself out. She has had an informal verbal warning and two private counseling sessions, each with formal reprimands. Lisa continues to run the machine hard and not use the safety shield. She still doesn't get it. *She is the unrepentant nonachiever.*

Step 4: Termination closes the loop. This time you will need quite a bit of preparation. There is no point for another reprimand because the two preceding reprimands and coaching haven't worked. You stop at her work station and ask for a meeting toward the end of her shift. You do not provide any real details for reasons that will become apparent.

At this step, termination is the only option, but you must call in some help here.

- First, you notify HR department of your intent and ask for a witness from HR to be present at the meeting in the afternoon.
- You ask that accounting prepares Lisa's last check that will pay her for up to the end of her shift regardless of when she actually walks out. You hand it to her on the way to the door after the final meeting in your office.
- You notify plant security for an escort to the parking lot and off the property. You ask plant security to make sure any parking privileges or passes are revoked effective at the end of the shift today. She may not reenter the building or the parking lot after today.
- You notify your IT people that Lisa's access to the company's records and remote terminals are blocked for obvious reasons of sabotage. All passwords are nullified.
- You ask your operations manager to meet you after the termination meeting so he or she can recover any company-owned tools or PPE before Lisa actually leaves the premises. It is much too difficult to do this after she leaves the company premises.
- Your own company may have other procedures and policies governing termination. Be aware of what termination involves at your own organization.

The central point in step 4, termination, is that you have moved permanently beyond coaching and behavior shaping and more meetings. Those efforts clearly haven't worked. Ending the company's relationship with the destructive employee is the single option remaining. The leader must be steadfast and can't cave in at the last minute in step 4. Lisa has changed her status for the last time: She is classified as one of *Dr. Winn's terminated nonachievers.*

Progressive discipline is a deliberate process of applying behavior shaping through coaching and at the same time increasing the price and penalty of nonperformance. It protects both leaders and employees from open and public confrontations. Progressive discipline is a good way to keep other departments abreast of what you do: HR, plant security, accounting, IT, and operations all know what is going on. There are really no surprises once step 4 is reached.

We say in class all the time that we generally avoid negative reinforcement as a means to modify behavior, preferring to use some sort or material or nonmaterial rewarding preference to punishment. An experienced leader knows, however, that there are times, we hope few in number, when punishment is in the form of progressive discipline, when either a person "gets with the program or gets out." As my friend Andy Peters says, "they simply can't work for you any longer."

Rose McMurray advanced in her career to the top safety job at the Federal Motor Carrier Safety Administration. As assistant administrator, she was charged with policy development and explaining agency activities to Congress, and she was also in charge of lots of employees. While she is exceedingly charming and diplomatic in person, she also knows when enough is enough. She has written a very useful sidebar here on disciplining difficult employees. Rose says this:

> I spent my federal career in promoting road safety and worked with the finest in the country. As safety professionals, you recognize the work you need to do is constant and the challenges never ending. I can't think of a more noble way to spend a career.

Having difficult employees in an organization is certainly a leader's challenge and can require time and effort to address, but once these employees are channeled effectively as a result of the leader's active involvement, intervention, and monitoring, they are capable of becoming fully contributing members of the workforce.

Rose is a master leader, if there ever was one. Figure 16.1 shows what she says about disciplining difficult employees.

Disciplining difficult employees

"I am thankful for the difficult people in my life. They have shown me exactly who I don't want to be." Ms. Rose McMurray, former chief safety officer, U.S. Federal Motor Carrier Safety Administration.

The effective leader recognizes that it is his or her obligation to exploit and harness the talents of all human resources in an organization even when certain of those resources tax the leader's patience and tolerance. In particular, difficult people exist all around us—within our families, at the checkout line, in our workplace. The leader's challenge is to know how to address the difficult employee head on and channel the employee's energy into productive work rather than into unacceptable work behaviors that negatively affect their colleagues and which compromise an organization's effectiveness.

In this discussion, it is important to distinguish the difficult person from the high-maintenance person. In my 38-year federal career, 32 of which were spent in supervisory positions, I dealt with employees who were demanding personalities but whose contributions to the organization were unmatched. These were typically bright, ambitious, innovative individuals who regularly sought my time but who consistently delivered high-quality, competent work. These workers were high achievers accustomed to praise and positive feedback. By understanding what motivated them to continue to be high-level producers, it was easy to keep them satisfied. On the other hand, the difficult employee is the one who seemingly is never satisfied, publicly complains about his/her situation, and generally displays unacceptable organizational behaviors.

Figure 16.1 An expert's suggestions for disciplining difficult employees.

(Continued)

For me, the litmus test for distinguishing the two types of workers is a simple question: "Is my organization better or worse off if this employee were to leave for another company?" The difficult employee is often the one gladly shown the door. However, many difficult people are well worth efforts to remediate and redirect them and there are many effective measures for dealing with them.

There are many reasons workers may be difficult. Among them are low self-esteem, a feeling of being underappreciated, unhappiness off the job, a perceived lack of control and influence, and often, being a "poor fit" in meeting job demands, etc. Regardless of why, a leader must respect *all* employees, even the difficult ones, and search for the motivators that may be unique to encouraging each individual to produce to his or her full talent. The leader has to embrace his or her responsibility to establish a fair and equitable workplace and deal competently with anyone who jeopardizes that condition. People in the organization expect the leader to address the situation.

One of the most important roles of the leader is to communicate performance expectations compared with actual performance accomplishments for all of his or her subordinates. Employees deserve this regular feedback in order to improve or calibrate their work.

With the difficult person, the manager needs to discern if the issue is performance based or attitude or behavioral. In my experience, it is much easier to deal with performance problems since remedies are more apparent and straightforward (more training, education, mentoring, etc.). When the problem is conduct or behavior, the person's personality and attitude are involved, making it more sensitive and uncomfortable to discuss. Sometimes, the difference between performance and behavior is hard to separate. Basically, if the person is technically skilled and able to produce the work but is unwilling to do so in an organizationally acceptable way, the issue is generally considered behavioral. If the leader stays committed to helping people succeed, he or she will undertake whatever reasonable actions are necessary to achieve it. As well, your conscience is clear that you acted with the best possible intentions.

Critical to being effective at remediating the difficult person is to recognize the organizational players (supervisors, human resources, attorneys, EEO staff, etc.) who need to be involved at the front end of any planned intervention. These consultations are critical to receive advice as you plan your performance improvement discussion, as well as provide an early alert that the difficult person may seek them out if the employee believes your attempts at performance improvement are unwarranted.

Here are some of the ways to tackle a difficult person having performance/conduct failures:

1. *Let the employee know they are valued and their work is important to the organization.* Communicate that you believe they have the technical capacity to be successful but the unacceptable conduct compromises the employee's value.

2. *Communicate with the employee about job requirements and their personal skills and abilities.* The worker may feel underemployed or unchallenged. Describe the current performance you are seeing from them and ask the employee what he or she believes may explain the unacceptable performance. Be prepared for the employee to vehemently disagree with your point of view. It is important to listen but to remain firm in your position. Remind the employee that he or she occupies a job that is important to be performed in a satisfactory way. By choosing to exceed expectations rather than underperforming, remind the employee that he or she demonstrates his or her readiness for a more demanding position and that you may be useful in helping them attain the next level.

Figure 16.1 (Continued) An expert's suggestions for disciplining difficult employees.

(*Continued*)

3. *Describe specific instances where the employee's conduct was inappropriate and the effect that behavior had on others and the organization's image.* Anticipate that the person will become defensive and will attempt to explain away the behavior. Hold your ground and re-play for them what would have been a more acceptable behavior. This is coaching an employee to a better outcome the next time a similar situation arises.

4. *Be certain that the employee receives a clear and unambiguous description of the problem(s) and your insistence that his or her attention to improving is non-negotiable.* By describing the norms of acceptable behavior, the individual should be able to see that the vast majority of his or her colleagues possess and exhibit conformance to these norms. As above, describe consequences if your expectations are not met.

5. *Engage the employee in a problem-solving discussion about whether certain assignments or situations would improve his or her job satisfaction.* Again, as a leader, remember you are trying to squeeze the best out of all of your resources. By involving the individual in finding solutions, the employee may see that your motives are honorable and sincere and you have a genuine commitment to seeing the employee succeed. Be clear that consequences exist if the employee does not improve or if he/she decides to ignore your expectations.

6. *Ask the employee what you can do to help them meet standards.* Try to determine what explains their attitudes and in what way you can offer help. For example, if the employee blames the team he or she is on as contributing to the issue, consider assigning him or her to a different team and see if the behavior improves. Remember, the employee may actually be less "difficult" but more unhappy with his or her current situation. In most cases, if the employee truly is difficult, he or she will have continued problems with other staff on other teams. This opportunity, however, allows the leader to determine whether the behavior is chronic or whether the organization is the "issue."

7. *Consider partnering the difficult person with a worker who is particularly adept at the "missing" skill(s).* For example, if the difficult behavior is a lack of tact in dealing with others, try placing the employee on a team with more skilled, tactful colleagues and urge the employee to observe and mimic these desired skills.

With all employee issues that require a leader's intervention, it is important to document your discussion, laying out the specific behaviors requiring improvement; describe specific interventions management is offering to help the employee succeed; and indicate the range of possible consequences (suspension, removal, etc.) if performance does not improve. Be sure to involve the employee in the terms of the performance improvement plan. This step will clearly establish expectations and will avoid future disagreement about what constitutes satisfactory performance. Ensure that regular, scheduled performance sessions are included to assess progress.

Rose McMurray

Figure 16.1 (Continued) An expert's suggestions for disciplining difficult employees.

Summary of the issue of discipline

In this chapter, I discussed a variety of methods for leaders to discipline nonachieving employees. These methods provide due process and accountability, but also give the employee a chance at redemption through coaching. In the long run of things, we'd all rather have an employee stop being a nonachiever and stay with the company: at a minimum, you save training and experience costs, and the terminated employee loses seniority. But since engineering and safety leaders are squarely in business ventures where people can, and do, get hurt, the leadership has to toe the line on substandard performance, and do it in an equitable, rational way. Progressive discipline is one option to offer stepwise, increasingly severe behavior shaping, and Rose McMurray provides a litany of helpful advice for a leader to mold behavior including providing unambiguous documentation of the substandard performance and being unflinchingly clear about exactly which behaviors are not up to par, yet still offering redemption opportunities along the way. The best kind of leader is always fair to employees, even to a fault, but steadfastly firm when the employee still chooses not to perform.

section four

Fine-tuning leadership applications

chapter seventeen

Organizational protocol for safety and engineering professionals
A brief introduction

Business protocol is the generally accepted way of acting at work. It is the rough equivalent of business manners. I placed the chapter on business protocol at the end of my book because it is material least founded in empirical research. Rather, business protocol is based almost entirely on accumulated wisdom and experience, and even though it dates back centuries, if not millennia, we won't encounter control groups or randomized experimental groups in the midst of this material. I haven't tried to provide any sort of comprehensive coverage of protocol because good materials abound, and I'd never be able to cover it all, anyway. I have selected topics carefully that certainly will apply to future safety professionals and engineers. Even though I encourage you to read elsewhere about business protocol, the topics I present are crucial to getting your career off to a good start. Pay attention.

I can hear you saying in the back of your mind, "Please, spare me a chapter on business protocol." Admit it. But these topics of protocol are extremely important for young safety professionals and engineers so you don't make embarrassing mistakes early in your career.

Here's what Will Rogers, the famous American humorist of the 1920s and 1930s, has said: "Good judgment comes from experience, and a lot of that comes from bad judgment." Let's try to avoid the major pitfalls, shall we?

The major topics I have selected include how to dress the part of a business professional, communicating and displaying confidence through nonverbal cues, office behavior and customs, regional affect and international nuances, business symbolism such as displaying the flag, and special occasions such as funerals and affairs of state.

If it makes the reader feel better, consider this: I get thank you cards fairly regularly from my students precisely because we discussed this material in class and *because they have applied it*. They talk to me about MBWA; they thank me for learning about progressive discipline.

Even though we're at the end of the book, don't give up on me now. I'll try to soften this with humor when I can, but this is really pretty serious stuff.

Even though there is not much actual research on office protocol, and recognizing that my thank you cards represent only conjectural, or anecdotal, evidence, there are many resources on business protocol—just as there are for leadership. I am drawing material from *Business Etiquette and Protocol* (2001) by Carole Bennett. I am also drawing from an old stand-by, Dale Carnegie's blockbuster book on business protocol, *How to Win Friends and Influence People* (1936). This bestseller has seen dozens of reprintings in dozens of languages. Honestly, reading the Carnegie book can still make me stop and take a deep breath because it is so very insightful. Good business manners do not go out of style.

Again, this isn't an exhaustive coverage of any of these topics, but only a discussion of some obvious mistakes to avoid. Let's start with first impressions.

Dressing the part

In my introductory graduate class, I see an even split between those who have decided to dress for success and those who retained the "undergraduate look" with a backward baseball cap and generally scruffy look about them. I tell them they need to think of themselves as "preprofessionals" right away because the ramp-up time is incredibly short. Here are some tips.

You may have never thought about having a special clothing store, for either gender, but I recommend it, and I also suggest finding this store soon after graduation. Every town has one, and they don't need to be filled with designer labels. The important point here is that, first, the local sales people know what styles are being worn at the office and for evening dress. Second, a good sales person gets to know you and what you like or don't care for.

Salespeople at good men's stores are professionals, the same as you. They won't steer you wrong and they won't let you make terrible mistakes. You can also try on clothes for fit, and often the alterations are usually free. Online stores don't make these offers, and even if they do, you'll wait a couple of weeks and hope that the clothes you need on Monday will arrive and fit on Saturday afternoon. Good luck with that.

In our region, the guys have Men's Wearhouse or Brooks Brothers, the latter a little pricier than the former. I have cultivated a relationship with my own sales guy and he always remembers where I work, where my kids go to school, and my basic styles. He also steers me to some sales opportunities. If he's busy with someone else, I come back or wait for him. He knows what kind of ties I like and he will even lay out a shirt, tie, and trousers combination to see if I like it. I am glad to pay a little more for that kind of service, and so should younger professionals.

Khakis and polo shirts are OK for casual wear, but consider starting with a dress shirt and tie. You'd never wear sandals or shorts to work for any reason. Some places allow it, sure, but let's start by shooting for the top.

If you are unsure where to start looking for a men's store, ask somebody local you trust who dresses well. Dressing in the Southwest, say, Albuquerque, is different from dressing for the Southeast, say Atlanta, and different again for the Northeast, say, Boston. Remember my basic rule: You need to look at least 10 percent better than everybody else, all the time, every day. Why? You never get a second chance to make a first impression.

Maybe I'm too picky in choosing work clothes, but I would avoid many department stores, and surely the big box stores. This is because they stock shirts and trousers and dresses and women's styles in only small, medium, and large and not half-inch sizes like the finer stores do. You get a better fit at the better stores and the service is unapproachable.

For women, Macy's, Nordstrom, and Elder-Beerman are a cut above the big box stores and have the same advantages I described for men. You get personal service and you get a larger variety of sizes, and again, alterations are often free.

Maria Dillon, a professional horsewoman and graphic designer in her own right, brought to my attention the things for women to remember about professional dress (Figure 17.1).

And to be fair and totally politically correct, I need a similar list for men. Until I get an official list, in Figure 17.2 are some suggestions from a local employee relations consultant for what men should never wear to work.

If you want to hit the ground running and travel light, men should buy a good pair of dark gray trousers, a couple of good nonprint cotton shirts, and a blue blazer, tie optional, with dark or ox blood loafers. This outfit is easily mixed, packs small to the plane, and projects "casual but professional" on day 1. You can easily switch to hard hat and work boots when the time calls for them.

Things women should never wear to work

Crocs, Uggs, fanny packs, scrunchies, sweatshirts and sweatpants, footless leggings and spandex pants, leather pants, face tattoos, velour tracksuits, message tees, too much skin, heavy or no makeup, what you wore yesterday, sequins, flip-flops, gobs of jewelry, sports jerseys, hats, long fingernails, wrinkled or torn clothing, a wallet chain, glitter, sunglasses inside, all-over animal print, short shorts, or a mini-skirt

Courtesy of Baxter, K., Hennings, A., Handley A., *Excelle, Networking for the Career Minded Woman*, 2007.

Figure 17.1 Dressing tips for women.

Things men should never wear to work

Ball caps or hard hats worn indoors; shorts; tank tops; tee-shirts with offensive humor; sandals or flip-flops; sweatpants; sports jerseys; sleeveless shirts; faded, torn, or jeans with holes (no matter how much they cost); droopy pants; excessive jewelry; body piercings; anything Goth; clothes that are too tight or too big; any visible tattoos

Figure 17.2 Dressing tips for men.

Bennett (2001) makes some good points for dressing regardless of gender. I have modified her list a bit, but I attribute these ideas to her experience:

- Be consistent with styles. If you're a suit person, be consistently a suit person. If you're a golf shirt person, make sure your supply is constantly refreshed. Unless you are out on a construction site, I would avoid faded jeans, period.
- Better to buy a shirt or blouse one size larger than one size too small. Get help from professional sales people who know how to fit your own body shape.
- Both genders are increasingly full figured. "Don't wear risqué clothing," says Bennett. "It is unprofessional to see a man's chest or biceps or belly peeking out from a shirt. Likewise, it is totally unprofessional to see a woman's cleavage or bare shoulders or too much bare leg. It may get attention—but not the kind of attention for long-term professional respect.
- If you plan to go someplace casual after work, bring extra clothes and change at the office.

I admit that this is a personal concern, but I promise you, it's real: *Shoes for both genders make a big first impression, good or bad.* Whether you are wearing canvas or leather, they need to be clean and fresh looking every day. Even if you wear steel or composite-toe boots, they need to be looking spiffy every day. Here's how.

Even if you have a shoe kit at home, you'll need one for the office eventually (Figure 17.3). The shoe kit should have paste-polish or liquid in an applicator. You'll need an old toothbrush to get the soles cleaned out and another brush to work in the polish. You ought to also have a bottle of heel dressing (in an emergency, heel dressing is the shiniest thing you can apply and it's quick, even on toes of shoes or boots). If you need leather boots, you can add to your kit a bottle of preservative that will also increase the shoes' resistance to water. Having wet feet on a construction jobsite in January makes for a very long day.

Figure 17.3 My very own 35-year-old shoe shine kit. It isn't pretty, but among other things, it has helped me prepare for testimony to Congress.

Second, if you use knee-length rubber or leather boots for work, you're not immune from the need for good-looking footwear. You'll need a small can of furniture polish or even WD-40 (that's right) and an old rag to keep them clean and presentable. Don't let the overspray get on your tile floor because it is really slippery. Spray these over a rug or just do it all outside.

Some odds and ends. An old tube of Chapstick makes a good lubricant for zippers on boots or rainwear. Have a set of thin rubber boots available on those days when you have to go out in rain or snow. Have a fold-up umbrella for emergency use and keep one in your car and also in your. For regular use or for guests, get a full-sized umbrella.

Neither men nor women should wear hats, caps, or headgear of any type when indoors. Carry them under your left arm or leave them at the door, likewise for sunglasses perched up on your head inside the building—please don't.

Office communication and behavior

Let's begin this sub-topic with something that has yet to be covered in any college class I have ever encountered, the office memorandum. I had to learn this the hard way: by making embarrassing mistakes. I'll try to spare you my shame.

The memo is a one- or two-page interoffice communication that is formal in nature but brief in composition. It often represents a summary of an issue and will often go to a permanent file, so preparation of the memo is a crucial window to your own professionalism. We use e-mails for most all other communications, but formal memos still hold a lot of power. Here's a minor clarification. One such communication is a "memorandum"; two or more are "memoranda."

I am suggesting Figure 17.4 as a model for setting up your own memoranda. Mine is not the only prototype available but is easily adaptable, and

---

MEMORANDUM

To: Mr. Guido Sarducci, Manager, Physical Plant, Exco

From: Gary Winn, Manager, Construction Safety and Health, Wyeco

Re: Parking space outside our building

Date: November 30, 2012

Guido:

As you probably recall in our discussion last week in your office, my department has doubled in size in the last year, and we are running out of spaces for my workers to park at your gravel lot. We are finding ourselves parking near the gate and near the fences which makes the travel lanes very tight. There have already been two minor incidents, and I'd like to avoid more of these.

Can you help me find more parking spaces for the members of my department during the final three months we are here on our temporary assignment? I anticipate we'll need about 20 total spaces, and I have some ideas about where we might find new spaces.

Let me know at good time to come by and discuss this. As usual, I am copying Fred Mork on this memorandum since he is the manager of the physical plant and parking facilities, and we discussed needing his input as we work together to solve this problem.

I appreciate your consideration of our needs.

GLW/gw

Cc: F. Mork, Operations Manager

Bc: M. Green

Figure 17.4 A simple memorandum with emphasis on format: Content will change, but the format remains.

until you create your own, mine is a good place to begin. I wish somebody had told me about the preparation of memoranda decades ago. I could have been better prepared on Day 1.

There is a lot going on in this sample memo. Notice that "memorandum" is in bold, all capitals and centered, while the other salutations are flush left. This is a pretty common style.

When I operated my small business and before I became a professor, I had no secretary. I had a special Word file called "memo-all," and it consisted of everything needed to get a memorandum started without the actual people, date, or message installed. It saved me a lot of time.

A memo is right to the point, not personal in any way. It is not funny or unbusiness-like. There is good reason for brief and business-like memoranda. A memo is usually a page, sometimes two pages, but when it is more than that, it becomes a letter or a novel.

Memos like these are sometimes pulled out of the file years—even decades—later to show, in the case of Figure 17.4 that my department actually did ask for additional parking spaces in case there is a question about our "due diligence."

Notice also at the bottom is a "Cc" for "courtesy copy," which means that everyone else knows who received a copy. Your own file copy may have the "Bcc" on it, which signifies "blind copy," and in this case, only you specify who else will see the memo.

I don't put memoranda on company letterhead. I recommend plain white paper for memoranda. Notice also the full-block style with no indentation for paragraph: very business-like and clean.

You put your initials right across your name in the salutation (second line) to indicate that this is the final copy. I use blue ink on memoranda and letters. In fact, I suggest using blue ink pretty exclusively at the office so you will always know which the original version is. Some companies require blue ink and nothing else.

Finally, you will note that my own signature line is not really a signature at all. It is my initials (all three) in capital letters and a "gw" in lower case letters, which indicates that I wrote the memo. Had my administrative assistant used my draft or e-mail draft for final preparation, the signature line would read "glw/mo."

Now that you know the basic format of a memorandum, adapt it to your own style or your company's style and use it consistently.

Some things to think about concerning business communications

Regarding letterhead, please be aware that there are some real liabilities to using letterhead for anything but bona-fide office communication. Don't use it to correspond with your kids' soccer coach, and don't use it to write a letter to the newspaper editor. The reasons here are simple, really. The use of letterhead is a tacit endorsement of your position by the company whose letter is represented. Think about it for a minute.

Did your company really endorse you asking for your kid's soccer number next summer to be the same as your old one when you used to play? Did your company really endorse you chastising the city for litter still on the streets after the Christmas parade? I seriously doubt it, but using letterhead gives the impression that your organization endorses your position. To be sure, ask your Human Resources professionals about company policy on use of letterhead. When in doubt, don't use company letterhead for anything but company business.

Another mistake to avoid is slipping your private correspondence into the organization's stamp machine and mail. "I didn't have a stamp," you say. Not only is it not ethical, but trust me, you don't want the organization's mail or accounting people reading the returned letters you wrote to your old college sweetheart and passed off as company mail. (Some time when we meet at a conference somewhere, ask me how I know this). Anyway, a small roll of stamps is a good investment in your own privacy.

Some things to know about office behavior

If your boss has an "open-door" policy, there are some things to know about what it really means. It does mean you can come to him or her with a problem that has cropped up and needs a word of advice. It does not mean come and talk about football or, God forbid, golf with a cup of coffee. Those meetings should brief and to the point. That's the formal message of open door.

In my own experience, the informal message of the boss's open door is as follows:

- Don't ever bring food or expect to stay past five minutes. If you need longer than that, make an appointment even if it's a few days away.
- Don't become the "entertainment committee" with jokes and endless small talk about the ball game last weekend. Trust me: Your boss doesn't care. For that matter, don't be the entertainment committee in anybody else's office. Don't wander the halls looking for somebody to listen to you tell about the deer you shot.
- Don't be the person who leaves a soft drink can in your boss's office. And remember this, if you leave a spit cup in a coworker's office, I will find you, I promise, and I will FedEx the cup back to you.
- When you get to your boss's office, knock in the prescribed way from Chapter 4 (two fairly loud knocks on the door or door frame even if the door is open) and don't sit down until you are invited to do so.
- Take your padfolio and carry it with you in the method I prescribed also in Chapter 4. Remember that carrying your padfolio under your

left arm leaves your right hand free to shake hands. Even if you are left handed, the custom in the United States is to shake hands this way.

- Take notes and follow up pretty quickly with a brief memo or a brief e-mail. The follow-up is as much for your memory as the boss's. A follow-up memo isn't always necessary, but it's easy and it's a good habit to develop.
- Nobody smokes inside anymore. Same goes for chewing tobacco, dipping, or taking a famous spit cup, although a water bottle seems appropriate, if odd to me.
- The boss's closed door means "come back later." Don't knock.

Try to be organized and prepared for an open-door meeting, in fact, any meeting. That means going back to your recap memo or notes and summarizing what you discussed then. This gets you ready to discuss the names of the contractors, job applicants, or the product line just to be sure you are current with it. Preparation means taking a few minutes ahead of time to think of the two or three things you really want to ask in your open-door meeting and to test the questions in your mind. And if you say, "I know you're busy so I'll just be a minute" and maybe don't even sit down, the boss knows you're sincere about you valuing her time. You'll be more welcome the next time.

Even in a phone call that's been planned ahead of time, jot down two or three items you want to discuss before you call or before the call gets to you. In days long gone, my offices always had administrative assistants to screen the calls and give you ten seconds to prepare and refresh. Now we don't have call screening, but we do have the person's name on the screen on most office phones. After you see who is calling, allow the phone caller an extra ring while you quickly prepare in your mind.

Some years ago, I started making a note of the telephone caller's name on my ever-handy yellow pad; this is just to remind me whose name to say when you're ready to hang up. Looking at the caller's first name in the closing seconds of a call prevents a silly but embarrassing slip-up. When the caller fails to call you by your name but you remember his or hers, you score some protocol points for being prepared.

Some things to know about making introductions

If we are honest with ourselves, I think we'd admit that we are uncomfortable making introductions to other people in a group, especially when we have international guests and we just know we're going to pronounce the name incorrectly.

We might be uncomfortable doing it, but taking the lead with introductions and a handshake shows nonverbal leadership among that group.

After the introductions are finished, I often sense the relief that some-body, even if not me, took the initiative.

I think it's better to make an honest effort and try your best with difficult names than ignore protocol and act as if that person has no name. I also think that those rules we learned in grade school about making introductions still work. I'll use Bennett's example verbatim. It's succinct and it works.

"When you introduce two people, look first to the person you con-sider to be most important. Say that person's name first, followed by "I would like you to meet..." Then look at the person being introduced and reverse the order. How you make introductions of people infers who you consider to be more important."

It helps during the introduction to add a pertinent comment about the person to get the conversation going. Note in the following example how the brief comments are added to the end of each introductory line. This provides a starting point for conversation between the individuals being introduced.

Introducing your boss to your visiting sister: "Dr. Smith, I would like you to meet my sister, Jan Edwards, who is going to have lunch with me today. Jan, this is the president of our college, Dr. Marie Smith."

It's a simple procedure that puts you squarely in charge of the imme-diate situation and shows people you know what you are doing. Making a little aside about being rival colleges, for example, is fine, but don't overdo the small talk here.

Somehow, I think the multitude of electronic technologies has made us so insular that doing basic introductions becomes very stiff. Too bad. These basic introductions, in my view, are a way to display leadership, good judgment, and an understanding of office protocol. All it takes is a bit of practice.

The next item of protocol is something I have learned myself and prac-tice myself, but Dale Carnegie (2009) says it so much better than I could. In fact, I am humbled that I even came up with this on my own. Here is Carnegie's passage about how to interest people in making conversation after introductions are completed.

> Everyone who was a guest of Theodore Roosevelt was astonished at the range and diversity of his knowledge. Whether his visitor was a cowboy, a Rough Rider, a New York politician or a diplomat, Roosevelt knew what to say. And how was it done? The answer is simple. Whenever Roosevelt expected a visitor, he sat up late the night before, reading up on the subject in which he knew his guest was par-ticularly interested. For Roosevelt knew, as all lead-ers know, that the royal road to a person's heart is to talk about the things he or she treasures most.

Protocol at your first engineering or safety conference

For the last decade and a half, all of the conferences I have attended follow the same format. Member conferences are roughly half trade show and half professional presentations with important guest speakers. You attend the interesting presentations and hear the high-ranking speakers and then attend the trade show to fill in the dead time. There will be the occasional evening meal, but most meals are on your own, and often extend well into the evening hours.

Conference dress will be influenced by the area of the country. Southwestern states are pretty casual affairs, while northeastern states are a bit more formal. You can't go wrong dressing up just a little beyond what you think the norm will be. Dressing down a little bit will cause people to remember you. Remember my rule about dressing 10 percent above what you expect to be the average.

And you don't want to be the young professional who didn't take a tie, who wore a ball cap to a conference, who took a spit cup into the meeting room, or who drank too much later in the evening.

OK. It's time to stop for a minute and discuss something important. This may be really important to your future and your career. A word to the wise here: Alcohol and late-night conferences are a sure invitation to inappropriate behavior, and the younger the person, the higher the probability of risk. I have had reports back to me about things I can't even mention here involving not only young people but also faculty, as despicable as that sounds. Please don't be that person. Make sure you are with somebody who probably has better sense that you do.

Driving is not the only risk. Don't allow yourself to be in a position where improper interpersonal things can happen under the influence of alcohol. Think of them ahead of time and avoid getting stuck in awkward situations or being alone on your walk back to the hotel or even in your hotel corridor. Try to leave all of the bad behavior in your undergraduate years.

What about presenting a conference paper yourself?

This has to be one of the best feelings ever: to present a report on data you collected yourself to support a theory or provide a conclusion to an exposure assessment. A publication is a real career builder for non-academics. The paper presentation and writing process go roughly like this:

- A "call for papers" is issued a year before the conference.
- You write a 150–200-word abstract, sometimes "blind," in which names are removed in order for the peer reviewers to ignore affiliations.

- You abstract is peer reviewed and then accepted or rejected.
- If your abstract is accepted, the full paper will be due in approximately three months for a final blind peer review. The paper formats are very strict but nothing that can't be overcome with time and effort.
- If the final paper passes peer review, it goes to publication in the proceedings, usually on CD but occasionally in hard copy and bound, and occasionally on line.
- You and any coauthors attend the conference and present a summary of the problem, methods, results, and conclusions in a half hour time block. In the last five or six years, and in two member associations where I publish regularly, papers that are published in proceedings are required to also be presented in person.

Your audience is probably going to be friendly and you'll see peers and networking associates in the audience. But either way, you have to practice the presentation like you are still in college and test the equipment ahead of time. Handouts are the norm, and good-color handouts make a lasting impression. Ask friends for potentially damaging questions.

The first time is all about butterflies and anxiety. With preparation, you'll be fine (Figure 17.5).

International nuance for young professionals

The days are long gone for a new engineer taking a job and staying local for 25 years. New professionals will be expected to travel almost immediately and increasingly overseas.

Americans travelling abroad are saddled with a wide variety of stereotypes: We're loud, we're rude, we eat too much, and we tip too little. Some of these are true, but only because the U.S. visitor didn't investigate whether he or she was going to step on toes acting that way. A little research certainly helps.

Bennett (2001) comes to our rescue with some words of wisdom about international travel.

> In general, try not to be too friendly too soon. Do not call people by their first names until they have given their permission. Resist the American behavior of quick informality. [People in] other countries take longer to "warm up" to people, and they generally observe a greater formality. Be patient when building trust in new relationships. For instance, when meeting someone from Great Britain for the first time, you would not ask what his or her occupation is.

Note to self, 40 years later...

Not all paper presentations are given to friendly audiences. Sometimes the audiences are hostile, and you have to be ready for that. For my very first paper presentation, I was 26 years old and I flew out to Vancouver, British Columbia, with just a suitcase and a naïve smile. I was sure that I didn't need to prepare anything other than making copies of my paper for my first conference.

As it turned out, I was presenting a research manuscript on a very politically-charged subject to about a hundred emergency room doctors 2500 miles away from my Midwest comfort zone. They were lying in wait for "fresh meat."

They had prepared a list of pointed questions for me, and it took me by surprise when I was put on the spot, again and again. I answered the best I could, sticking to my data, but I don't think they were really satisfied because I avoided their political predilections. I lived to tell the tale, but I vowed to work harder at what might happen next time. I was both shaken up and angry at the same time.

Right before the next time came to present that kind of paper, I sat down with an attorney friend of mine, Gene, and he made sure he peppered me with every rude, leading, untoward, and unexpected question that a crafty Alabama lawyer could think of. It's a valuable lesson to learn as a young professional, even now. I recommend thinking about what could possibly go wrong at your presentation, from failures of your electronic files, burned out projector bulbs, power failures, and being slammed with rude questions.

You have to be ready.

Dr. Winn

Figure 17.5 Not all presentations go according to plan.

- In most parts of the United States, people generally are not very generous with the terms *please* and *thank you*. Additionally, the terms, *yes, sir* and *yes, ma'am* are rarely heard, except in the southern states. Because of this neglectful habit, Americans are generally perceived to be impolite. Most people from other countries use "please" and "thank you" generously, as well as the courtesy titles. When entertaining people from other countries, you may want to sprinkle these terms into your vocabulary.
- Treat international guests as very important people in your life. Escort them to a seat or at least indicate where they might be seated; offer to hang up their coat. Show them that you really care about their well-being.
- If an international visitor hands you a business card at your first business meeting, accept it with both hands, and scan it immediately for vital information. Then lay the card in front of you on the table. Others may consider it demeaning if you put the card directly into your pocket without looking at it first; it may also be considered impolite to write on someone's card in his or her presence.

- Sports are always one safe topic. Mention someone from their country who did well in the Olympics or a winning soccer team. Not golf.
- Speak slowly and enunciate each syllable. Those who have learned English from a textbook or in a classroom environment generally comprehend individual words (but not always contractions or colloquialisms). Saying "I do not know" in four clear syllables is easier to comprehend than "I don't know" and certainly better than "I dunno."
- Try to avoid American slang, which has become so much a part of our daily lives. For example, imagine how a person learning English would comprehend this monologue: "I'll be doggoned. He drives me nuts. Who gives a darned if we spend an hour grazing this buffet table. Tell him to bug off. I'll give him a piece of my mind when I return." How could you provide a section of your brain to another person?

Professionals of tomorrow will be working with a wide variety of people from across the globe. I have asked a few of my international graduate students to suggest to young professionals in the United States a bit of their own culture.

Kuwait

Major Butti Al-Butti is in the Kuwaiti National Guard and he studied safety management in the United States before returning to the Middle East. Here is what "The Major" says about doing business in Kuwait.

> Kuwaitis are hospitable; however, it is significant to act according to their cultural norms especially if you are doing business in Kuwait. They prefer to do business with those whom they have a personal relationship and spending a great deal of time on the getting-to-know-you process. You must observe patience since impatience is viewed as a criticism of the culture and you should dress and present yourself well. They respect education, so carefully mention if you have an advanced degree, especially if it is from a prestigious university.
>
> When working as a safety engineer, meetings are preferred in the early evening, but when dealing with government officials, it should be in the morning since they are restricted to a 6-hour day. Meetings may be interrupted with prayer times and are generally not private unless there is a need to

discuss matters confidentially. They are event rather than time driven, so expect that the event of getting together is more important than the timeliness in the meeting or the outcome. They are of hierarchical society and most companies are structured around the family. Decisions usually come from the top after determining the consensus of various stakeholders. Decisions are reached slowly, and if you rush things, you will give offense and risk your business relationship. They are smart negotiators who are especially interested in price, so do not use high-pressure sales tactics. Repeating your points indicates you are telling the truth. Proposals and contracts are kept simple, negotiations are done in English, and contracts are written in both English and Arabic versions; the Arabic will be the one followed.

Thailand

Mr. Narupon Thankiul has an undergraduate degree in IT engineering from King Mongkut's Institute of Technology, Thailand (Figure 17.6). After that, and before he came to study in the United States, he built his career at Advance Info Services Company, the largest telecommunication company in Thailand, as a senior engineer in the network operation department. While perusing his master's degree in safety management at WVU, he took internship in security, safety, health, and environmental policy and standard development division with PTT, Inc., the biggest Thai petroleum company. Here is what Pon suggests about doing business in Thailand.

Don't talk about politics if you're on business in Thailand. Most people are separated into two big groups called Red shirt and Yellow shirt. Even though the situation is better than two years ago, there are still several political assemblies and sometimes violence erupts. It's best not to attend these rallies.

Don't say or do anything to disrespect the Thai King. Thai people have the highest respect for our King, so if you do anything like that, you will get some pushback. Also be aware that there is a law to protect the King from scorn, which may cause you to be fined or even imprisoned. Don't even make jokes.

Figure 17.6 Pon (left) at his graduation in 2012 from WVU's Safety Management program with Dr. Winn.

Gambling, drugs, and prostitution are illegal in Thailand no matter what you have heard. Most Thai people are Buddhist. Therefore, you should know some common practices about Buddhism, such as a woman must not wear shorts or a sleeveless shirt when going to a religious place or that a woman is not allowed to touch a monk.

Thai people are kind and like to help others, but most Thai people are not good in English and tend to avoid talking to foreigners. If you can speak a few words in Thai, it would be great for the beginning of the conversation. That's reasonable.

You should know that some areas in southern Thailand still have terrorist activities, especially at the border of Thailand and Malaysia. Therefore, they use martial law in some areas. Although these areas are not tourist places or business areas, you should prepare yourselves and make sure that you are always in the safer areas.

Normally, working hours for government organizations are 8:30 a.m. to 4:30 p.m., and for private companies, it is 8:30 a.m. to 5:30 p.m. However, if you want to do business with a government organization, you should go earlier because the pace is slow and some people get off early.

You can use an English-language contract to go to court in Thailand, but the judge might ask for the Thai version of your case, which has to be translated by a certified person or organization.

Some businesses are not allowed for foreigners to do, such as telecommunication. You can be a partner or stockholder, but the owner or the major shareholder has to be operated by Thai person or a Thai company.

Generally, I think doing business in Thailand is the same as anywhere else overseas. You should know the laws and regulations in that business you want to do. If you have some connections, especially if you are doing business with the government, it would be good for your business dealings. Thailand still has a lot of corruption, but it does not mean that corruption is legal in Thailand. Mostly, Thai people welcome foreigners to invest in our country because it will generate many jobs and money.

Actually, I should mention that most Thai people think that Nation, Buddhism, and King are the most important things in Thailand. Therefore, anything that refers or represents those three things is also important, too. For example, the Thai national flag is an important symbol of the country. At a recent Lady Gaga concert in Thailand, one of her performers had a Thai national flag hung on a motorcycle and a part of the flag touched on the stage. After that picture was broadcast, there were a lot of criticisms about the inappropriate show, and finally, she had to apologize to the Thai people.

India

Nitya Narayanan recently completed her master's program in Safety Management at WVU. Prior to starting her graduate studies, Nitya was employed in the IT industry as a user experience designer for more than 12 years both in India and in the United States. Nitya is originally from Chennai, a populous and diverse metropolitan city on the south eastern coast of India, and she brings an insider perspective on conducting business in India, growing up in an entrepreneur family. She makes these key points.

> India has attracted and maintained the attention of global investors in recent years in spite of the global economic slowdown because of its growth potential and optimistic expectations for its future. But the business climate in India is still a tricky proposition, and anyone venturing to pursue such an option should always remember these social norms, in addition to conventional business guidelines, to be successful.
>
> - Indians value education a lot, especially a graduate degree from the United States. They will hold you in very high regard and it might give you an added advantage in marketing your product.
> - It is always good to listen to others discussing politics or religion, but it would serve well to avoid contributing to such conversations. India is still quite conservative and there is significant emphasis on morality, respect, beliefs, and faith.
> - Indians are predominantly religious and typically start any task with a short prayer and remembering God. Different states in India have different regional languages that official business is conducted in, apart from English and Hindi, which is the national language.
> - Although the handshake is common when meeting people, Indians also use the "Namasté" gesture. Namasté is where the palms are brought together at chest level with a slight bow of the head. Using the Namasté is a sign of your understanding of Indian etiquette.

- It is very important to show respect to anyone, especially the elders. Traditionally, Indians greet each other by standing up and using the handshake or the Namasté gesture.
- Be sure to receive anything and give anything with your right hand, as using your left hand is considered disrespectful.
- When entering a meeting room you must always approach and greet the most senior person first.
- Unfortunately, corruption is common and widespread in all levels of government. So, there is no use in complaining but it would be better to deal with this issue delicately. Here is where hiring a street smart local consultant would definitely help.

Nigeria

Akinbolusire Oluwaseun is a master's degree candidate in the Safety Management Program, coming to the United States from Nigeria. She has an undergraduate degree in agriculture and plans to return to Nigeria to blend these fields.

I urge your professionals coming to Nigeria to be sensitive to local customs and needs and to study these ahead of the travel. Companies must ensure that their travelers understand the values, needs, and behavior patterns of the local consumer. Research reveals that low-income markets often have unreliable sources of income. Erratic cash flow, for example, affects packaging, marketing, and shipping strategies in Nigeria.

- Regulatory requirements: Any new entrants to African markets must familiarize themselves with local requirements. It cannot be assumed that goods that meet the quality standards of a company's home country will automatically meet the regulatory requirements of other African countries.
- Educate yourself about Africa from sources other than the Discovery Channel. Two good choices are contacting the embassy directly,

and a good second choice is contacting the local university; often, there is a chapter of students from the local area who will be willing to "key you in" on local customs.

- Visit the location several times if you intend to establish a company or an outlet. Each time, travel out and talk to others who hire or trade with local people. What are the labor strengths and weaknesses? Sometimes, you will be surprised that there are small but sophisticated high-tech companies who can fill your organization's needs. Ask!
- Follow up on every arrangement to make sure that conditions and terms are fully understood and are being implemented. Don't leave these things to chance. If you translate as is often required, make sure your translator knows about legal, accounting, and contract language.
- Cell phone use is expanding very rapidly and almost everybody knows how to use a smart phone.
- Be mindful of the three Ps of African businesses: pensiveness, patience, and perseverance. Pensiveness demands the use of common sense. Success demands a lot of patience. Above all, you must persevere and be persistent but patient. The pace is slower, plain and simple.
- Lucrative as it may be, the African market is not without pitfalls. Keep your guard up and insist on cash transactions as far as possible. Establish direct contacts with your business counterparts in Africa by participating in trade fairs and exhibitions. Be wary of intermediaries.

Angola

Zinga Martinelli completed her WVU graduate degree after working now for large multinational oil company. She has an undergraduate engineering degree and she has worked in her native Africa for more than 10 years. She reported the following common ethical situations in Angola in Southern Africa.

- Whenever a person enters a room (office, medical) where other people are already in the room, the newcomer must take the initiative first and greet the people, generally saying "Bom dia/Boa tarte" (hi). It would be regarded as very impolite to just enter the room or the local and sit down.
- It is also regarded as offensive to sit in an office (school or office) with feet on the table or showing other people around the bottom of your shoes.
- Talking with your superior or Angolan public officials with your hands in pocket is highly disrespectful. I heard anecdotal reports one time that an American working in Angola in a multinational oil company was delayed entry in the country for days during the immigration process just for that. The act was perceived as rude.
- In Angola, pictures may not be taken in public places, including the airport, without the proper permit procedure. Disregarding this aspect may get you in trouble. It does not matter if you are taking pictures of your friends or something else. And remember, there may not be any sign or warning saying so.
- When it comes to business, physical appearance sends a strong message. People are generally expected to use not only professional work attire but also formal outfits, that is, a clean men's or lady's suit, but no bright color. In some public places, doors may open or close due to overdressing or dressing "loud" in a manner that attracts undue attention. I once heard that a well-established businessman was stopped when entering an office while his driver was kindly invited in. The driver was wearing a trouser-suit while he was not! Go figure, but be aware it could happen to people who do not explore local customs ahead of time.
- When a person is invited for dinner/lunch in a restaurant, it is not a common thing to share the bill or have the invited person pay for his part of the meal. Even though it is assumed

that the person inviting will pay the bill, the invitee usually feels a need to repay this kindness by treating the one other person the same way or even better.

- In some parts of Angola, respect due to hierarchy is very important. Commonly, subordinates or younger people are expected not to stare at superiors or older people when greeting them. In fact, in these situations, looking at the person's eyes may be regarded as challenging or disrespecting to the person's authority. In the United States, the custom is opposite: avoiding eye contacts when greeting people may be regarded as shyness or lack of transparency.

Summary of rules about international travel

Without question, young safety professionals and engineers are going to travel soon after they are hired, first in the continental United States and, soon, overseas. The traveler will need to know all about necessary immunizations, visas, and passports. And unlike 30 years ago, extensive travel is often a key part of the job of almost every safety professional and project engineer, and as such, you'll need to brush up on that Spanish or Italian you thought you'd never use from high school.

The probability of international travel goes up as the organization's size increases. We looked at business customs and examples of local protocol in Thailand, Angola, India, Kuwait, and Nigeria. Almost all of our international students recommended checking out local rules of etiquette via the country's embassy or a local university. Since there is no Wikipedia for international travel, the young professional going on his or her first business trip is advised to be polite, unassuming, and dressing low key at all times. All of us must be aware that the stereotypes of rude Americans being vulgar, loud, and dressing poorly arrived in their country long before you did. Let's prove them wrong.

Business symbolism: Honoring the American flag

I don't get to claim much connection to icons in American history, but I'm claiming this connection because it's related to the American flag. My current house was built on a Revolutionary War grant given ultimately to the Ross family. Searching the deed and available documents years back, I found that the Ross family historian from the 1800s claimed a familial connection to Betsy Ross—remember her as the maker of the first U.S. flag for George Washington? The claim may be fatuous, but it's a good story to tell my visitors.

Americans are traditionally proud of flag-oriented stories and traditionally proud of their flags. Unfortunately, not a lot of people know that there are actually laws—sections of U.S. Code—governing flag use. Let's explore just a few of those because, increasingly, organizations want to show their support of local soldiers who are away in the National Guard or just to express their own patriotism. There are ways to fly and care for the U.S. flag the right way, and there are ways to do it wrong, too.

Volume (title) 4 of U.S. Code, Chapter 1, discusses the flag's dimensions, colors, and how it is constructed. Title 18, Chapter 33, Section 700, discusses criminal penalties for using the flag incorrectly on purpose, known as "flag desecration." Title 36 of the U.S. Code (note that this is in a different place in the Code), Chapter 3, pertains to patriotic customs and observances using the flag.

Trivia material: The words to the Pledge of Allegiance are contained in Title 4, Chapter 1, paragraph 4. Paragraph 6 tells us that the flag should be displayed on the 19 official days of the year, or when states declare so, or (more trivia material) on state birthdays, that is, their dates of admission to the union. Here are some useful bits of knowledge from the U.S. Code about displaying the flag.

- Paragraph 6a: It is the universal custom to display the flag only from sunrise to sunset on buildings and on stationary flagstaffs in the open. However, when a patriotic effect is desired, the flag may be displayed 24 hours a day if properly illuminated during the hours of darkness.
- Paragraph 7b: The flag should not be draped over a float or car but placed on a staff (a small pole).
- Paragraph 7c: No other flag or pennant should be placed above or to the right of the U.S. flag.
- Paragraph 7m: When used at half-staff (for funerals or official mourning periods), the flag should be raised to the top of the staff briefly and then lowered to half-staff position.
- Paragraph 8a: The union of the flag (blue field) should never be down, except in case of dire emergency or distress.
- Paragraph 8c: The flag should never be carried horizontally, but always aloft and free.
- Paragraph 9: When the flag is passing in a parade or in review, or when playing the national anthem of the United States, all persons present in uniform should render the military salute. Members of the Armed Forces and veterans who are present but not in uniform may render the military salute. All other persons present should face the flag and stand at attention with their right hand over the heart or,

if applicable, remove their headdress with their right hand and hold it at the left shoulder, the hand being over the heart. Citizens of other countries present should stand at attention.

- Paragraph 8j: No part of the flag should ever be used as a costume or athletic uniform. However, a flag patch may be affixed to the uniform of military personnel, firemen, policemen, and members of patriotic organizations. The flag represents a living country and is itself considered a living thing.

Every once in a while, the flag becomes a real symbol of freedom, and even then, the flag must be used properly. In 1964, students at Balboa High School in the Panama Canal Zone objected when Panamanian students wanted to place their country's flag above the American flag in front of the U.S. high school on United States-held soil. The American students in the photo in Figure 17.7 prevented that. It was a pretty serious week—over

Figure 17.7 On January 11, 1964, students at Balboa High School in the Panama Canal Zone protect the American flag, touching off a week of riots.

20 people died in the ensuing riot. My best friend Terry is the guy with his hand raised and I am in the background somewhere.

Later, as a member of the high school's ROTC unit, I raised the American flag there myself a number of times on the pole in the photo, and believe me, I was proud to do it. You should be proud to raise your organization's own flag.

A bit about funerals

We've all been to funerals, and I provided a brief chapter about the "death event" that sometimes catches employers and even professionals off guard. But even early in your career, you may be called upon in an official capacity to organize pallbearers or deliver a eulogy. I found a few bits of business etiquette related to funeral behavior on *The Etiquette Page*, a Canadian blog that's really pretty good. Here are the bits verbatim that may be useful if you've never been faced with the organizational aspect of a funeral. I recommend reading the full page for a fuller treatment.

> *What to say to the bereaved:* Expressing sympathy to someone in deep mourning can be difficult. It's best to keep it short and simple. On arrival, greet the family and briefly offer your condolences.
>
> - Be specific when offering any help. Offer to help with childcare, make dinner, or run errands, for example.
> - Avoid claiming that you know how someone feels. Simply let them know that you're thinking of them.
> - Don't bring up spirituality.
>
> *What to do* if you are asked to serve as a pallbearer or usher at the service, even if the request comes at the last minute. In sensitivity to the family, you will agree immediately to help out. Here's what to expect:
>
> - *Pallbearers:* Pallbearers carry or escort the coffin to the burial place. During the service, you will sit at the front, just behind the family. If you are asked to be a pallbearer and are not comfortable carrying the coffin, you may be able to escort it instead.
> - *Ushers:* Ushers help escort mourners to their seats before the service. Always try to seat those with the closest relationship to the

deceased nearest to the front. Ushers themselves can sit wherever they choose once the ceremony starts.

What to do if you are asked to deliver the eulogy: You may be asked to give a eulogy at the service. If you are not comfortable doing so (or too upset), it's perfectly okay to decline. If you do decide to say a few words, keep these tips in mind:

- A eulogy is just 2–10 minutes long.
- Plan and write out what you are going to say before you arrive (you will be anxious and notes will help steady you).
- Have someone review your words beforehand.
- Share how you knew the deceased, and don't speak only about your relationship with him or her.
- Emphasize the deceased's best qualities.

Source: *The Etiquette Page Blog*

Summary of office protocol and procedures

This overview of office protocol was not intended to be exhaustive. There are other people much more expert about office protocol than I am. What I wanted to provide here was merely a snapshot of what you'll need to know within the first year on the job. My goal was merely to make you aware in a general way. This includes some basic advice about how to dress if you are unsure, knocking on your boss's door, how to write a simple memorandum, some international travelling rules, and courtesies about the flag. Leaders know these things, and first impressions are important. I am only interested in having you make fewer mistakes than I did when I was your age...

chapter eighteen

Summary of this book's key concepts

In the 17 preceding chapters, we have made a remarkable journey in the pursuit of "jump starting" the early careers of safety professionals and engineers and moving them through the ranks to prepare them to become leaders in their organizations. When looking for a textbook for my own use in class, I found nothing to fill the gap for young people entering or currently employed in safety or engineering who wish to become and to train leaders, so I decided to write my own.

I present the chapters here in the same order that a young professional would encounter them.

I introduced the need for the book based on my extensive talks and empirical research with my students over the years in both safety and engineering. I had a hunch that the demographics of our engineering and safety students had changed: While students are altruistic and unselfish to a sometimes scary degree, both sets of students had missed important opportunities to work, to travel, and to learn about the world around what it means to be a professional and not just an employee. Most important, because most young people have worked little or not at all before or during college, they have missed other opportunities to see how genuine leaders work and act and how they train their own subordinate leaders.

I used a lot of military examples because it turns out that we have similar, if not identical missions and we have used similar, if not identical organizational research to support leader development, like values assessment, identifying toxic leadership, and pointing out the need for experiential training. And unlike selling insurance for a living, when leaders in safety or engineering make bad decisions, their employees and others can suffer, even die. We all have to be right the first time.

Entering a profession

My goal was to help ensure career success by building some competence in preprofessionalizing skills and knowledge, to encourage reading and outside learning in a very wide context, to study important canons of ethical conduct by the organizations our students will eventually join. I shared some simple wisdom that I have gained over the years, and I tried to make a point with them. If I seemed self-aggrandizing, that was not my goal; on the contrary, by now, the reader should realize that I abhor it.

Becoming a leader

After this long buildup, I introduced leadership in crisis and noncrisis models. I talk about when good leadership goes bad and what to do about it. I discuss psychological and physiological stressors and ways to overcome them. Because safety and engineering are more diverse and global every day, I figured that a primer on these topics, backed with research or interviews, was a good idea for those entering the professions or those entering recently.

To me, maybe the most important chapter is how to create a system of leadership and leader development in the microenvironment even when the people upstairs in the big offices don't seem to care. I called this material developing leaders in a "depleted" environment. It only takes the persistence and grit of a motivated individual to examine our motives and to begin with an agreement for an honor code. That's what leaders do, even when nobody is looking. And maybe it seems out-of-date to have an honor code, but when I introduce the concept in class, it gets attention like almost no other topic ever does: Students seem eager to be shown ways to set their own moral compasses.

Most classroom "training" in most industries today is actually education. Instead of doing more PowerPoints, authentic leaders will opt to do a kind of time-honored experiential training which brings hands-on experience to leader development. Not every employee will need it or want it, but when we train in mission-critical tasks (rescue and fire protection or fall protection, for example), subordinates will benefit from following the algorithm presented in the chapter on the modified model for experiential training. I also show that it is cost-effective and provides opportunities for current and future leaders to learn from each other.

In some high-risk occupations, the probability of having an occupational fatality is higher than in others. Consequently, we need to be prepared for the "death event." As I prepared this text, I got a lot of pushback for bringing attention to the fact that, in these businesses, we are sometimes the focus of negative publicity. "Why bring attention to it?" some reviewers asked. The better to be prepared, I say, and I hope you agree.

Leaders are called upon to perform difficult tasks, sometimes including handing fatal injuries or employee deaths; they will need to apply discipline in a measured and repeatable way. We discuss progressive discipline and we learn from Rose McMurray's experience handling difficult employees.

In a chapter on gender, we examined some gender-role research, and then I invited some young professional engineers and safety professionals to describe their experiences. There still seems to be gender bias out there, but the numbers of women engineers and safety professionals are growing, if slowly. Men and women apply equally effective safety and engineering programs, but they will probably do it in different ways.

The crippling effects of poor organizational morale can be identified and offset by working proactively with employees and their families. Current research suggests that positive attitudes and behaviors can be learned and that effective strategies can be developed to protect people and resources.

Finally, and as a wrap-up, I gathered material to suggest what young people are going to need to know about office communication, dress, acting in international contexts, and even the best methods to be used in business symbolism such as patriotic display of the American flag.

As Carl Heinlein declared in the Introduction to the book, the future is incredibly bright and rich for students interested in safety or engineering. Safety students and engineers apply what they have learned about math, science, humanities, and behavioral science and apply these skills directly to social problems in which they can be of help.

Industry is crying out for leaders in these professions. Remember that in the foreseeable decade, we can expect the creation of two safety jobs for every person entering the workforce and that, for engineers, the top 7 of 10 jobs offered in the United States are to engineers, and salaries and job satisfaction are higher than in other fields. Neither are bad ways to start a career.

My goal has been to ensure that the young people who are entering or are currently employed in these professions and who want to transition smoothly from student to professional to leader take as few missteps as possible, know the value of acting proactively, and make the best decisions possible to preserve and protect people and property.

I talked about it all for 25 years. It was time to write the book. I sincerely hope I have met your expectations. If you have comments, please contact me.

Bibliography

American Society of Safety Engineers. (2014). 1800 E. Oakton St., Des Plaines, IL 60018.

Andrew, M., McCanlies, E., Burchfiel, C., Charles, L., Hartley, T., Fekedulgen, D. et al. (2007). Hardiness and psychological distress in a cohort of police officers. *International Journal of Emergency Mental Health*, 137–147.

Ballard, G. (2005). *Servant Leadership: More Philosophy, Less Theory*. Bloomington, IN: AuthorHouse.

Banks, C. B. (2010, October 28). *How Companies Can Develop Critical Thinkers and Creative Leaders*. Retrieved 2012, from Harvard Business Review Blog Network: http://blogs.hbr.org/2010/10/how-companies-can-develop-crit/.

Barker, J. A. (1993). *Paradigms: The Business of Discovering the Future*. New York: Harper Business.

Barling, J. (2014). *The Science of Leadership: Lessons From Research for Organizational Leaders*. New York: Oxford University Press.

Barling, J., Weber, T., & Kelloway, K. E. (1996). Effects of transformational leadership on attitudinal and financial outcomes: A field experiment. *Journal of Applied Psychology*, 81(6), 827–832.

Baxter, K., Hennings, A., & Handley, A. (2009, May 26). *25 Things Women Should Never Wear to Work*. Retrieved 2014, from Excelle, Networking for the Career Minded Woman: http://excelle.monster.com/news/articles/3398-25-things-a-professional-woman-should-never-wear.

Bennett, C. (2001). *Business Etiquette and Protocol*. Cincinnati, OH: South-Western. Thomson Learning.

Bennis, W. (2002). The leadership advantage. In F. Hesselbein & R. Johnston, *On Mission and Leadership: A Leader to Leader Guide* (pp. 7–18). San Francisco: Jossey-Bass.

Blanchard, K. H., & Johnson, S. (2003). *The One Minute Manager*. New York: Morrow.

Blankenhagen, E. E., & Rozman, T. R. (2009, June). A training concept for heavy forces. Modernization. *Military Review*.

Board of Certified Safety Professionals. (2002). Code of Ethics.

Burke, K. (2014, June 18). GM whistleblower was told to "not find every problem," report says. *Automotive News*.

Card, O. S. (1991). *Ender's Game*. New York: Tor.

Carnegie, D. (2009). *How to Win Friends and Influence People*. New York: Simon & Schuster.

Chang, K. (2012, September 25). Bias persists for women of science, a study finds. *New York Times*, p. D1.

Collins, J. (2005, July 1). Level 5 leadership: The triumph of humility and fierce resolve. *Harvard Business Review*.

Collins, J. C. (2001). *Good to Great: Why Some Companies Make the Leap—And Others Don't*. New York: Harper Business.

Collins, J. C., & Hansen, M. T. (2011). *Great by Choice: Uncertainty, Chaos, and Luck: Why Some Thrive Despite Them All*. New York: Harper Collins.

Collins, J. C., & Porras, J. I. (1997). *Built to Last: Successful Habits of Visionary Companies*. New York: Harper Business.

Comprehensive Soldier Fitness. (2011). *America Psychologist Special Issue*.

Connelly, O. (2002). *On War and Leadership: The Words of Combat Commanders from Frederick the Great to Norman Schwarzkopf*. Princeton, NJ: Princeton University Press.

Conti, R. S. (2001). Responders to underground mine fires. *Proceedings in the Thirty-Second Annual Conference of the Institute on Mining Health, Safety and Research, Salt Lake City, UT*.

Coram, R. (2002). *Boyd: The Fighter Pilot Who Changed the Art of War*. Boston: Little, Brown.

Cornum, R., Matthews, M. D., & Seligman, M. E. (2011). Comprehensive soldier fitness: A vision for psychological resilience in the U.S. Army. *American Psychologist*, 1–3.

Corsini, R. J., Craighead, W. E., & Weiner, I. B. (2010). *The Corsini Encyclopedia of Psychology*. Hoboken, NJ: Wiley.

Covey, S. R. (1989). *The 7 Habits of Highly Effective People: Powerful Lessons in Personal Change*. New York: Simon & Schuster.

Crandall, D. (2007). *Leadership Lessons From West Point*. San Francisco: Jossey-Bass.

Dawson, M. M., & Overfield, J. A. (2006). Plagiarism: Do students know what it is? *Bioscience Education e-Journal*. doi:3108/beej.8.1.

Deming, W. E. (2000). *Out of the Crisis*. Cambridge, MA: MIT Press.

Donnell, J. A., Aller, B. M., Alley, M., & Kedrowicz, A. A. (2010). Why industry says that engineering graduates have poor communications skills: What the literature says. *American Society for Engineering Education, 2011 Annual Conference and Exposition, Vancouver, British Columbia*.

Eagly, A. H., & Karau, S. J. (2002). Role congruity theory of prejudice toward female leaders. *Psychological Review*, 573–598.

Eckert, A. W. (1996). *That Dark and Bloody River: Chronicles of the Ohio River Valley*. New York: Bantam Books.

Ewing, Phillip. David Petraeus plea could pave way for comeback. *Politico*. March 3, 2015. http://www.politico.com/story/2015/03/david-petraeus-plea-115723.html.

Fisher, B. (1974). *Small Group Decision Making: Communication and the Group*. New York: McGraw-Hill.

Goldbaum, E. (2012). *Police Officer Stress Creates Significant Health Risks Compared to General Population, Study Finds*. University at Buffalo News Center. http://www.buffalo.edu/news/releases/2012/07/13532.html#sthash.LEXPwWqH.dpuf.

Goldratt, E. M. (1999). *Theory of Constraints: What Is This Thing Called and How Should It Be Implemented?* Croton-on-Hudson, NY: North River.

Goodwin, D. K. (1998, Summer). Lessons of presidential leadership. *Leader to Leader*, pp. 23–30.

Goodwin, D. K. (2005). *Team of Rivals: The Political Genius of Abraham Lincoln*. New York: Simon & Schuster.

Greenleaf, R. K. (1977). *Servant Leadership: A Journey into the Nature of Legitimate Power and Greatness*. New York: Paulist Press.

Greenleaf, R. K., Spears, L. C., Covey, S. R., & Senge, P. M. (2002). *Servant Leadership: A Journey Into the Nature of Legitimate Power and Greatness*. (25th Anniversary ed.). New York: Paulist Press.

Hammond, G. T. (2001). *The Mind of War: John Boyd and American Security*. Washington, DC: Smithsonian Institution.

Hansen, M. T. (2009). *Collaboration: How Leaders Avoid the Traps, Create Unity, and Reap Big Results*. Boston: Harvard Business.

Harms, P. D., Herian, K., & Vanhove, L. (2013, April). Report #4: Evaluation of resilience training and mental and behavioral outcomes. *The Comprehensive Soldier and Family Fitness Program Evaluation*, (4), 19.

Harris, C. E., Pritchard, M. S., & Rabins, M. J. (2005). *Engineering Ethics: Concepts and Cases*. Belmont, CA: Thomson/Wadsworth.

Hawking, S. (1998). *A Brief History of Time*. New York: Bantam.

Henderson, G. F., & Wolseley, G. (1936). *Stonewall Jackson and the American Civil War*. London and New York: Longmans, Green.

Herzberg, F., Mausner, B., & Snyderman, B. B. (1959). *The Motivation to Work* (2nd ed.). New York: John Wiley.

Hesselbein, F., & Johnston, R. (2002). *On Mission and Leadership: A Leader to Leader Guide*. San Francisco: Jossey-Bass.

Hesselbein, F., & Shinsecki, E. K. (2004). *Be, Know, Do: Leadership the Army Way: Adapted from the Official Army Leadership Manual*. San Francisco: Jossey-Bass.

Hindle, T. (2008, November 28). *Guru: Geert Hofstede*. Retrieved 2012, from The Economist: http://www.economist.com/node/12669307.

Hofstede, G. J., & Minkov, M. (2010). *Cultures and Organizations: Software of the Mind* (3rd ed.). New York: McGraw-Hill.

Hogan, R., & Curphy, G. (2008). *Leadership Matters: Values and Dysfunctional Dispositions*. Retrieved 2012, from Curphy Consulting: http://www.leadership keynote.net/index.htm.

Holtzapple, M. T., & Reece, W. D. (2006). *Concepts in Engineering*. Dubuque, IA: McGraw-Hill.

House, J. S., Wells, J. A., Landerman, L. R., McMichael, A. J., & Kaplan, B. H. (1979). Occupational stress and health among factory workers. *Journal of Health and Social Behavior*, 139.

Howe, N., & Strauss, W. (2004). *Millennials Rising: The Next Great Generation* (3rd ed.). New York: Vintage Books.

Huber, T. (2007, January 1). 12 months after Sago, coal industry still changing. *Washington Post*.

Hughes, R. L., Ginnett, R. C., & Curphy, G. J. (2009). *Leadership: Enhancing the Lessons of Experience*. Boston: McGraw-Hill Irwin.

James, A. P., Stotz, C. M., Masich, A. E., & Halaas, D. F. (2005). *Drums in the Forest: Decision at the Forks*. Pittsburgh: Historical Society of Western Pennsylvania, University of Pittsburgh.

Kaltman, A. (1998). *Cigars, Whiskey & Winning: Leadership Lessons from General Ulysses S. Grant.* Paramus, NJ: Prentice Hall.

Kelloway, K. (2006). *Handbook of Workplace Violence.* New York: SAGE.

Keenan, V., Kerr, W., & Sherman, W. (1951). Psychological climate and accidents in an automotive plant. *Journal of Applied Psychology,* 108–111.

Keith, K. M. (1998). *The Case for Servant Leadership.* Westfield, IN: Greenleaf Center for Servant Leadership.

Kelloway, K. E. (2003). Workplace safety. In P. Warr, *Psychology in the Workplace.* London: Penguin Books.

Kelloway, K. E., & Barling, J. (2000, September). Knowledge work as organizational behavior. *International Journal of Management Review,* 287–304.

Klann, G. (2003). *Crisis Leadership: Using Military Lessons, Organizational Experiences, and the Power of Influence to Lessen the Impact of Chaos on the People You Lead.* Greensboro, NC: Center for Creative Leadership.

Kolditz, C. T. (2009, February 6). *Why the Military Produces Great Leaders.* Retrieved 2012, from Harvard Business Review Blog Network: http://blogs.hbr .org/2009/02/why-the-military-produces-grea/.

Kolditz, T. A. (2007). *In Extremis Leadership: Leading as if Your Life Depended on It.* San Francisco: Jossey-Bass.

Kotter, J. (1990). *A Force for Change: How Leadership Differs From Management.* New York: Free Press.

Kouses, J. M., & Posner, B. Z. (2007). *The Leadership Challenge.* San Francisco: Jossey-Bass.

Kuhn, T. S. (1968). *The Structure of Scientific Revolutions.* Chicago: University of Chicago Press.

Leoncioni, P. (2002). The trouble with humility. In F. Hesselbein & R. Johnston, *On Mission and Leadership: A Leader to Leader Guide* (pp. 41–52). San Francisco: Jossey-Bass.

Lindenburg, C., Gendorp, S., & Reiskin, H. (1993). Empirical evidence for the social stress model of substance abuse. *Research in Nursing and Health,* 351–362.

Lipkin, N. A., & Perrymore, A. J. (2009). *Y in the Workplace: Managing the "Me First" Generation.* Franklin Lakes, NJ: Career Press.

Long, A. L., & Schindler, S. (1995). *Memoirs of Robert E. Lee.* New York: Crescent.

Lovelace, J. B. (2012). *PL300 Military Leadership Course Reader.* New York: XanEdu (Jossey-Bass).

Mager, R. R., & Pipe, P. (1984). *Analyzing Performance Problems.* Belmont, CA: Lake Publishing Co.

Marrella, L. (2009). *In Search of Ethics: Conversations with Men and Women of Character* (3rd ed.). Sanford, FL: DC Press.

Matthews, P. J., & Lester, P. B. (2011). *Leadership in Dangerous Situations: A Handbook for the Armed Forces, Emergency Services, and First Responders.* Annapolis, MD: Naval Institute.

Maxwell, J. C. (1998). *The 21 Irrefutable Laws of Leadership: Follow Them and People Will Follow You.* Nashville, TN: Thomas Nelson Publishers.

McAdams, M. T. (2011). National Assessment of the Occupational Safety and Health Workforce. Wesstat, Rockville, MD. Prepared under contract for the National Institute of Occupational Safety and Health.

McPherson, J. M. (1988). *Battle Cry of Freedom: The Civil War Era.* New York: Oxford University Press.

The Myers & Briggs Foundation. (2014). Retrieved from myers-briggs.org.

Myer, A. (1968). *Once an Eagle.* New York: Holt, Rinehart and Winston.

National Science Foundation. (2012, January). *Science and Engineering Indicators.* Retrieved 2014, from National Science Board: http://nsf.gov/statistics /seind12/.

National Society of Professional Engineers. (1997). Code of Ethics.

NCTI. (2014). Retrieved from realcolors.org.

Ness, J., Jablonski-Kaye, D., Obigt, I., & Lam, D. M. (2011). Understanding and managing stress. In P. J. Sweeney, M. D. Matthews, & P. B. Lester, *Leadership in Dangerous Situations: A Handbook for the Armed Forces, Emergency Services, and First Responders* (pp. 40–59). Annapolis, MD: Naval Institute.

NIOSH. (1999). *DHHS (NIOSH) Publication Number 99-101: Stress...At Work.* Cincinnati, OH: National Institute of Occupational Safety and Health.

NIOSH. (2011, October 20). *National Assessment of the Occupational Safety and Health Workforce.* Retrieved 2012, from cdc.gov: http://www.cdc.gov/niosh/osh workforce/.

Northouse, P. G. (2013). *Leadership: Theory and Practice.* Thousand Oaks, CA: SAGE.

Nowack, K. (1997). Personality inventories: The next generation. *Performance in Practice,* American Society of Training and Development, Winter 1996/97.

O'Brien, J. (2001). *At Home in the Heart of Appalachia.* New York: Knopf.

Osland, J. S., Kolb, D. A., Rubin, I. M., & Turner, M. E. (1995). *Organizational Behavior: An Experiential Approach.* Upper Saddle River, NJ: Prentice Hall.

Ott, J. S. (1989). *The Organizational Culture Perspective.* Chicago: Dorsey Press.

Padilla, A., Hogan, R., & Kaiser, R. (2007). The toxic triangle: Destructive leaders, susceptible followers, and conducive environments. *The Leadership Quarterly,* 176–194.

Page, M. (2014). *Everyday Etiquette.* Retrieved 2014, from etiquettepage.com: http://etiquettepage.com/everyday-etiquette.

Peters, T. J. (1982). *In Search of Excellence: Lessons From America's Best-Run Companies.* New York: Harper & Row.

Peterson, D. (2003). *Techniques of Safety Management: A Systems Approach.* Des Plains, IL: American Society of Safety Engineers.

Pittenger, D. J. (1993, Fall). Measuring the MBTI and coming up short. *Journal of Career Planning and Placement.*

Prosser, S. (2010). *Servant Leadership: More Philosophy, Less Theory.* Westfield, IN: Greenleaf Center for Servant Leadership.

Quenk, N. L. (1999). *Essentials of Myers-Briggs Type Indicator Assessment.* New York: John Wiley.

Real Colors. (2005). National Curriculum and Training Institute, Inc., Phoenix, AZ.

Reed, B., Midberry, C., Ortiz, R., Redding, J., & Toole, J. (2011). Morale: The essential intangible. In P. J. Sweeney, M. D. Matthews, & P. B. Lester, *Leadership in Dangerous Situations: A Handbook for the Armed Forces, Emergency Services, and First Responders* (pp. 202–217). Annapolis, MD: Naval Institute.

Reid, P. C. (1990). *Well Made in America: Lessons from Harley-Davidson on Being the Best.* New York: McGraw-Hill.

Rhodes, J., & Jason, L. (1987). *The Social Stress Model of Alcohol and Other Drug Abuse: A Basis for Comprehensive, Community-Based Prevention.* Washington, DC: U.S. Department of Health and Human Services.

Rhodes, J., & Jason, L. (1998). *Preventing Substance Abuse among Children and Adolescents.* Elmsford, NY: Pergamon Press.

Rogers, C. R., & Freiberg, H. J. (1994). *Freedom to Learn* (3rd ed.). New York: Merrill.

Rozman, T. R. (1991a, May–August). The mechanized infantry rifle company as a leadership academy. *Infantry*.

Rozman, T. R. (1991b, November). Maneuver and gunnery training for tomorrow. *Military Review*.

Rozman, T. R. (1992, Winter). Training support. *Army Trainer*.

Rozman, T. R. (1993, March). *Theater Level Training Strategy, Military Review*.

Rozman, T. R., & Saunders W. A. (1992, March–April). An exercise in leadership, *Infantry*.

Ruben, E., Sapienza, P., & Zingales, L. (2013). How stereotypes impair women's careers in science. *Proceedings of the National Academy of Sciences of the United States of America*, 111(12), 4403–4408.

Sandberg, S. (2010, December). *Why We Have Too Few Women Leaders*. Retrieved June 2014, from Ted.com: https://www.ted.com/talks/sheryl _sandberg_why_we_have_too_few_women_leaders.

Sandberg, S. (2013). *Lean In: Women, Work, and the Will to Lead*. New York: Knopf.

Schein, E. H. (2004). *Organizational Culture and Leadership*. San Francisco: Jossey-Bass.

Schreiner, E. (2012). *Consequences of Plagiarism for Professionals*. Retrieved 2012, from eHow.com: http://www.ehow.com/list_6017871_consequences-plagiarism -professionals.html.

Seligman, M. E. (2011). *Flourish: A Visionary New Understanding of Happiness and Well-Being*. New York: Free Press.

Seppala, E. M. (2014). *Feeling It: Emotional Expertise for Happiness and Success*. Retrieved 2014, from Psychologytoday.com: http://www.psychologytoday .com/blog/feeling-it.

Shaara, M. (1974). *The Killer Angels; a Novel*. New York: McKay.

Sipe, J. W., & Frick, Dm.m (2009). *Seven Pillars of Servant Leadership: Practicing the Wisdom of Leading by Serving*. New York: Paulist Press.

Sontag, D. (2004, May 27). The struggle for Iraq: Interrogations. How a colonel risked his career by menacing a detainee and lost. *The New York Times*.

Stephan, E. A. (2011). *Thinking Like an Engineer: An Active Learning Approach*. Boston: Pearson.

Strife, M. L., Armour-Gemmen, M. G., & Hensel, R. A. (2012). Engineering a bridge to information literacy. *2012 American Society for Engineering Education Annual Meeting*. San Antonio, TX: American Society for Engineering Education.

Sunell, R. J., & Rozman, T. R. (1987). Future training with the armored family of vehicles, *Proceedings, 9th Interservice/Industry Training Systems Conference*, Washington, DC, November 30–December 2, 1987.

Sweeney, P. J., Matthews, M. D., & Lester, P. B. (2011). *Leadership in Dangerous Situations: A Handbook for the Armed Forces, Emergency Services, and First Responders* (pp. 202–217). Annapolis, MD: Naval Institute.

Tarrants, W. E. (1977). What is a safety professional? *American Industrial Hygiene Association Journal*, A4.

Taylor, M. (n.d.). *The West Point Honor System: Its Objectives and Procedures*. Retrieved 2012, from west-point.org: http://www.west-point.org/users /usma1983/40768/docs/taylor.html.

Tulgan, B. (2009). *Tulgan, Bruce*. San Francisco: Jossey-Bass.

Uldrich, J. (2005). *Soldier, Statesman, Peacemaker: Leadership Lessons from George C. Marshal*. New York: AMACOM.

Vandiver, F. E. (1957). *Mighty Stonewall*. New York: McGraw-Hill.

Violanti, J. M., Andrew, M., Burchfiel, C. M., Hartley, T. A., Charles, L. E., & Miller, D. B. (2007). Posttraumatic stress symptoms and cortisol patterns among police officers. *Policing: An International Journal of Police Strategies and Management*, 30, 169–188.

Waddle, S., & Abraham, K. (2002). *The Right Thing*. Nashville, TN: Integrity.

Webb, J. (1994). *Born Fighting*. New York: Broadway Books.

Winn, G., & Banks, B. (2014, January). Safety leadership: Insights from military research. *Professional Safety*.

Winn, G., Giles, B., & Heafey, M. (2012). Maximizing the value of internship opportunities for safety management graduate students in the Appalachian Region. *Proceedings of the 2012 ASSE Professional Development Conference*. Denver, CO: The American Society of Safety Engineers.

Winn, G., Hensel, R., Curtis, R., Taylor, L., & Cilento, E. (2010). An integrated approach to recruiting and retaining Appalachian engineering students. *American Journal of Engineering Education*, 2(2), 1–16.

Winn, G., Jones, D., & Bonk, C. (1992). Taking it to the streets: Helmet use and bicycle safety as components of inner city youth development. *Clinical Pediatrics*, 672–677.

Winn, G., Jones, D., & Bonk, C. (1993). Testing the social stress prevention model in an inner city day camp. *Transportation Research Record*, 106–116.

Winn, G., Winn, L., Hensel, R., & Curtis, R. (2009). Data driven comprehensive mentorship: How we are adapting the social-stress model of peer influence at West Virginia University. *American Society for Engineering Education Annual Conference Proceedings*. Austin, TX: ASEE.

Winn, G. L., Seaman, B., & Baldwin, J. C. (2004, December). Fall protection incentives in the construction industry: Literature review and field study. *International Journal of Occupational Safety and Ergonomics*, 10(1), 5–11.

Winn, G. L., Williams, A., & Heafey, M. (2013, June). A research-based curriculum in leadership, ethics and protocol for safety management and engineering students. *Proceedings of the American Society for Safety Engineers, Annual Conference*. Las Vegas, NV: The American Society of Safety Engineers.

Yeo, S. (2007). First-year university science and engineering students' understanding of plagiarism. *Higher Education Research and Development*, 26, 199–216.

Yoder, B. (2013). Engineering by the numbers. In A. S. Education, *2013 ASEE Profiles of Engineering and Engineering Technology Colleges*. Washington, DC: American Society for Engineering Education.

Zohar, D., Livne, Y., Tenne, O., Admi, H., & Donchin, Y. (2007). Nursing climate scale: A tool for measuring and improving safety. *Clinical Care Medicine*, 1312–1317.

Index